AF904591

ADVANCES IN PETROCHEMICALS

Edited by **Vivek Patel**

Advances in Petrochemicals
http://dx.doi.org/10.5772/59296
Edited by Vivek Patel

Contributors

Nasser Awwad, José Luis Velázquez-Ortega, Omar Reyes-Martínez, Mohamed Abdel-Aziz Younes, Elijah Adekunle Taiwo, Juan Pereira, Lei Zhang, Hangzhou Wang, Tong Qiu, Bingzhen Chen

Published by InTech

Janeza Trdine 9, 51000 Rijeka, Croatia

© The Editor(s) and the Author(s) 2015

The moral rights of the editor(s) and the author(s) have been asserted.
All rights to the book as a whole are reserved by InTech. The book as a whole (compilation) cannot be reproduced, distributed or used for commercial or non-commercial purposes without InTech's written permission. Enquiries concerning the use of the book should be directed to InTech's rights and permissions department (permissions@intechopen.com).
Violations are liable to prosecution under the governing Copyright Law.

Individual chapters of this publication are distributed under the terms of the Creative Commons Attribution 3.0 Unported License which permits commercial use, distribution and reproduction of the individual chapters, provided the original author(s) and source publication are appropriately acknowledged. More details and guidelines concerning content reuse and adaptation can be found at http://www.intechopen.com/copyright-policy.html.

Notice

Statements and opinions expressed in the chapters are these of the individual contributors and not necessarily those of the editors or publisher. No responsibility is accepted for the accuracy of information contained in the published chapters. The publisher assumes no responsibility for any damage or injury to persons or property arising out of the use of any materials, instructions, methods or ideas contained in the book.

Publishing Process Manager Iva Lipovic
Technical Editor InTech DTP team
Cover InTech Design team

First published October, 2015
Printed in Croatia

Additional hard copies can be obtained from orders@intechopen.com

Advances in Petrochemicals, Edited by Vivek Patel
p. cm.
ISBN 978-953-51-2176-3

We are IntechOpen, the world's leading publisher of Open Access books

Built by scientists, for scientists

3,800+
Open access books available

116,000+
International authors and editors

120M+
Downloads

151
Countries delivered to

Our authors are among the
Top 1%
most cited scientists

12.2%
Contributors from top 500 universities

WEB OF SCIENCE™

Selection of our books indexed in the Book Citation Index
in Web of Science™ Core Collection (BKCI)

Interested in publishing with us?
Contact book.department@intechopen.com

Numbers displayed above are based on latest data collected.
For more information visit www.intechopen.com

For EU product safety concerns:
IN TECH d.o.o., Prolaz Marije Krucifikse Kozulić 3, 51000 Rijeka, Croatia,
info@intechopen.com or visit our website at intechopen.com.

Meet the editor

Vivek Patel has a strong background in materials science, marketing, and management, and he graduated from Central Institute of Plastics Engineering and Technology (CIPET). Previously, he was associated with Lucintel, Sperry Plast Ltd., and major chemical MNCs. His areas of expertise are polymers, composites, chemicals, and nanotechnology as well as writing, reviewing, and editing technocommercial magazines such as *Modern Polymers & Plastics*, *Chemical World*, *Chemical Weekly*, *FRP Today*, *Popular Plastics*, and *Packaging*. In the past, his article on nanomaterials for Li-ion battery has been recognized by *Forbes* and *Times* of India. His innovation won Next Big Idea Business Competition, organized by Indian Institute of Management Bangalore (IIM-B), Intel and Department of Science and Technology (DST), Government of India, and showcased in UC Berkeley Business Plan Competition, USA. Mr. Patel also finalized in FICCI Innovation, Eureka, Power of Ideas, and Economic times' business idea competition.

Contents

Preface

The petrochemical industry is an important area in our pursuit of economic growth, employment generation, and basic needs. It is a huge field that encompasses many commercial petrochemical and polymer-enabled products. The objective of this book is to provide the critical insights on synthesis and processing, mechanism, technologies, and modeling and simulation in the petrochemical research domain to research scholars, academics, research scientists, science and engineering students, and industry professionals.

This book contains six chapters written by various distinguished authors. The selection of topics addressed and the examples, tables, and graphs used to illustrate them are governed, to a large extent, by the fact that this book is aimed primarily at the petroleum science and engineering technologist. This book is designed to help the reader, particularly students and researchers of petroleum science and engineering, to understand synthesis, processing, mechanics, and simulation of the petroleum processes. The content is as widely encompassing as possible for the petrochemical research field. The book is based on the series of chapters written by various distinguished authors and divided into three parts: namely, introduction, synthesis and processing, and modeling and simulation as well as six chapters, each one succinctly dealing with a specific focus on the petroleum domain.

Vivek Patel
SKO Centre for Knowledge Management of Nanoscience and Technology (CKMNT),
Vijayapuri Colony, Tarnaka,
Secunderabad, Telangana
India

Section 1

Introduction

Chapter 1

Cashew Nut Shell Oil — A Renewable and Reliable Petrochemical Feedstock

Elijah A. Taiwo

Additional information is available at the end of the chapter

http://dx.doi.org/10.5772/61096

Abstract

This study evaluates the occurrence of cashew plant, extraction processes and composition of cashew nut shell liquid (CNSL), modification and conversion processes, as well as environmental impact and controls of the liquid as petrochemical feedstock. The goal of this study is to bring the alternative usage of CNSL to the limelight and to mitigate the serious problems posed by the depleting petroleum reserves.

Cashew nut shell (CNS) was obtained fresh from the wild, pulverized and then extracted by pyrolysis and solvent extraction methods using *n*-hexane. The liquid obtained was characterized. Experimental results corroborated the fact that CNSL contains mainly phenolic compounds such as cardanol, cardol, anacardic acid and 6-methyl cardol. CNSL obtained by pyrolysis has a higher density with less moisture content, while solvent-extracted CNSL had higher contents of cardol and anacardic acid.

CNSL offers innumerable applications due to the phenolic nature of its constituents, with enshrined features for transformation into high-value specialty chemicals. Varied CNSL composition with varying extraction processes are a vantage opportunity with multiple application potentials as a valuable petrochemical feedstock.

Keywords: Cashew nut shell Liquid (CNSL), Petrochemical feedstock, Solvent extraction, Renewable resources, Characterization, Supercritical fluid

INTECH
open science | open minds

© 2015 The Author(s). Licensee InTech. This chapter is distributed under the terms of the Creative Commons Attribution License (http://creativecommons.org/licenses/by/3.0), which permits unrestricted use, distribution, and reproduction in any medium, provided the original work is properly cited.

1. Introduction

Cashew Nut Shell Oil (CNSO) is a versatile component of the Cashew fruits' nut. The oil which is a dark reddish-brown in colour is resident in a soft honeycomb shell, that is the pericarp of the nut. It is a natural resin that could serve as a valuable raw material for multiple applications. Rapid growth in the world population and increasing standard of living has overstretched petroleum resources as petrochemical feedstock. This among other factors, has culminated in the fast depletion of global petroleum reserves. Therefore, to maintain the standard of living and continuity of industrial sector which is paramount to human survival this decade and beyond, there is a need to find alternative sources of fuel and petrochemical feedstock. Cashew nut shell, a by-product of the cashew industry, is an embodiment of a useful chemical serving as a raw material for the petrochemical industry.

Cashew (*Anacardium Occidentale* L.), a well-known species of the Anacardiceae family [1], is a tropical plant (shrub) found within the region between 23°N and 23°S of the equator. It is a drought resistant tree crop grown successfully in areas with annual rainfall of 50–350 cm. Its height is above 12 m and has a spread of about 25 m. It has an extensive root system which makes it adaptable to a wide range of moisture levels and soil types [2]. On the commercial scale, it is best grown on well-drained, sandy loam soils. Today, it is an economically valuable tree crop in Nigeria, with evergreen leaves, all year round.

The cashew tree consists of the cashew nut fruit (which is a curved edible seed, housed in a honeycomb-like shell), the apple, leaf and bark. The fruit consists of an outer shell, inner shell and the kernel. The thickness of the cashew nut shell is about 1/8 in. (0.32 cm). The soft honeycomb matrix, in between the outer and inner shell, contains a dark brown liquid, which is known as cashew nut shell liquid (CNSL) [3].

The Cashew plant is of great economic significance to Nigeria and other tropical countries due to its valuable products. These products are utilized in food, medicine, chemical and allied industries. In addition to providing shade, the cashew tree is an embellished ornamental plant that suitably controls soil erosion.

Usually, it takes about 3-4 years from planting time before fruiting. Thereafter, the tree can live up to 40 years and more. The cashew fruit (figures 1 and 2) is unusual in comparison with other tree nuts since the nut is outside the fruit. The cashew apple which is about 10 cm long is an edible false fruit, attached to the externally born nut by a stem. Its color ranges from yellow to red, fibrous in nature, very juicy, sweet, pungent and high in vitamins A and C. In its raw state, cashew nut shell (CNS), which is leathery in nature, contains the vesicant oily liquid (CNSL). The shell is separated from the kernel by the testa which is a thin skin surrounding the kernel.

1.1. Growth and development of cashew in Nigeria

The world production of cashew crop, according to the Food and Agriculture Organization (FAO), was around 2.7 million tons per annum. The major raw cashew producing countries with their production figures in 2005 (as per the United Nation's Food and Agriculture

http://dx.doi.org/10.5772/61096

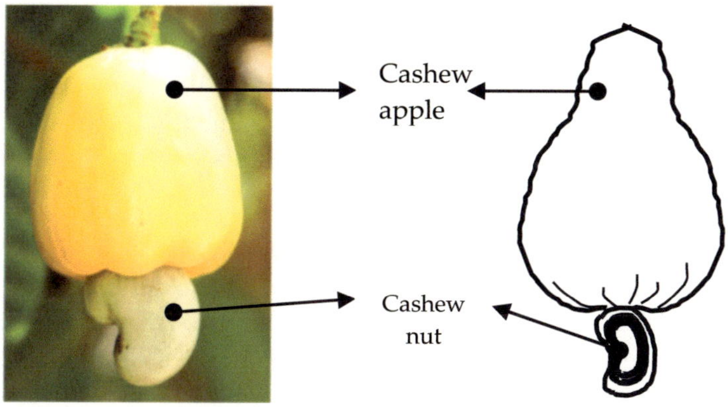

Figure 1. The Cashew fruit.

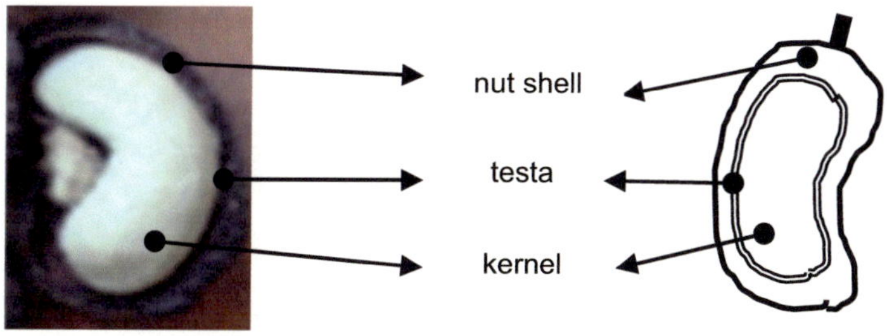

Figure 2. The cashew nut cross section

Organization) were Vietnam (960,800 tons), Nigeria (594,000 tons), India (460,000 tons), Brazil (147,629 tons) and Indonesia (122,000 tons). India ranks first in area utilized for cashew production, though its yields are relatively low. Collectively, Vietnam, India and Brazil account for more than 90% of all cashew kernel exports [4,5].

Cashew production in Nigeria dates back to the 15th century when it was introduced by Portuguese explorers [4,5]. The plant was then, purposefully planted for afforestation schemes in the now defunct Eastern Nigeria. Cashew became a popular commercial crop in 1953, when planted on a large scale principally for the nuts, afforestation and erosion prevention programmes in the escarpment areas of Dui, Mbala, Oghe, Oji, Isuochi and Kingie in Eastern Nigeria by the defunct Eastern Nigeria Development Corporation. In Western Nigeria, the first planting of cashew started in the 16th century at Agege in Lagos [4]. Commercial cultivation actually started in the 1950s at Iwo, Eruwa and upper Ogun in the defunct Western Nigeria by the then Western Nigeria Development Corporation.

Cashew was thereafter, introduced into the Middle Belt and Northern Nigeria from Eastern and Western Nigeria. These cashew nuts were mainly of the medium nut-size biotype, which according to ISO-6477 standards, fall within the W320 category. This cashew biotype therefore constitutes the main cashew species in Nigerian cashew plantations. This biotype, which was from the Asian continent, attracts low premium in the international market. However, Cocoa

Research Institute of Nigeria (CRIN) which was established in 1962, commenced research on the cultivation, uses and economy of cashew around 1972. The crop then rapidly spread to all agro-ecologies of Nigeria [6]. This brought about the cultivation for erosion control and the afforestation schemes of the Brazilian cashew biotype (jumbo nut-size), which is now being grown by cashew farmers [6]. This species mature within a year in contrast to the local wild varieties which take about 3 to 5 years to mature.

Today, cashew grows almost everywhere in Nigeria but its cultivation is concentrated primarily in the south and middle belt regions both in small holder farms and plantations [2]. Major cashew growing areas in Nigeria include Abia, Abuja, Anambra, Benue, Cross river, Enugu, Ebonyi, Ekiti, Kogi, Kwara, Kebbi, Nasarawa, Niger, Oyo, Osun, Ondo, Ogun, Sokoto, and Taraba states. High quality cashew nuts, suitable for export, are produced from the south west and south east regions of Nigeria. Figure 3 shows the spread of cashew production in Nigeria as published by National Cashew Association of Nigeria.

Establishment of the Premier Cashew Processing Factory in Oghe, Enugu State early in the 1980s marks the turning point in industrial processing of cashew in Nigeria. The primary objective of government then was to process the harvest from the government-owned 650 ha cashew plantation as well as to serve other small land holding farmers in and around the States of Kogi and Benue.

Figure 3. The spread of cashew cultivation in Nigeria

In 2002, Nigeria produced about 30,000 tons of cashew nut from the total holdings of 50,000 ha, which were mostly under small holdings. The average growth of cashew nut production increased to 55,000 tons in 2004. In 2010, Nigeria ranked third among the top 10 world cashew nut-producing countries with a total production of 594,000 tons [7]. This stride improved to second worldwide in 2011 with a production of 835,000 metric tons, constituting 19.5% of the world's production according to FAOSTAT data 2013 [7]. The world production of cashew nuts is presented in Figure 4.

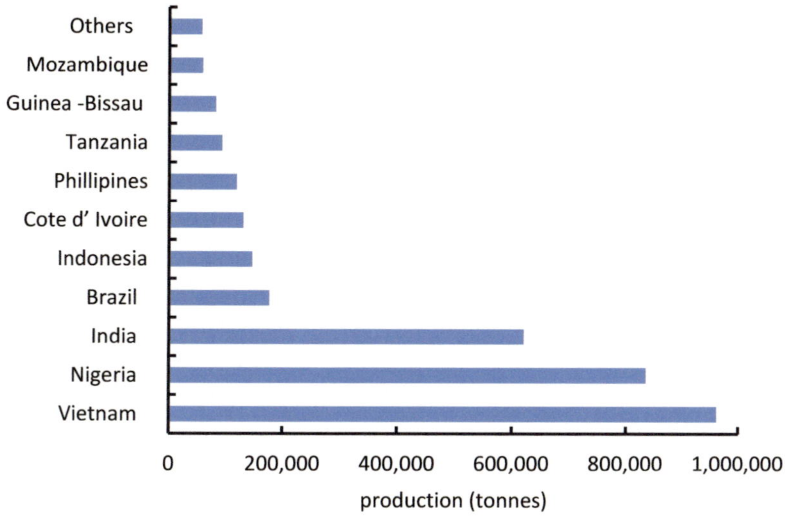

Figure 4. Main producer countries of cashew nuts

Recently, a cashew export programme was initiated in Lagos, Nigeria by the Nigerian Export Promotion Council (NEPC), the National Cashew Association of Nigeria (NCAN) and USAID/ Nigeria Expanded Trade and Transport (NEXTT). The move is to expand further the production of cashew nut for export purposes. The National Cashew Association of Nigeria (NCAN) is the umbrella body for the Nigerian cashew industry. It is actively involved in developing a strategy and action plan to improve the quality of raw nuts, to increase the coordination in the sector and to improve the business climate for processing.

India is a leading exporting country of cashew nuts. Its production volume stems from the skilled labour force and technology deployed at the processing stage. The popularity of cashew nuts arose from its applicability as natural vitamin pills [(balita.ph/.../cashew-nuts-gain-popularity-as-natures-vitamin-pill/)]. Cashew kernels on the other hand, have high protein content (about 19.5%), which accounted for its use in place of soybean meal in broiler chickens [8], whereas the nuts serve a major source of alkenyl phenolic compounds [9].

1.2. Processing

Basic physical operations before extraction are required on the samples of cashew nuts to ensure a high degree of purity and the quality of the product. These operations constitute the basic pretreatment on samples and include washing, drying, shelling and size reduction. Sometimes, washing may involve the use of detergents to remove likely contaminants.

Drying is purposely to make the nuts moisture-free. Both sun- and oven-drying have been found effective. Size reduction creates a better contacting surface area for the shell and solvent to enhance removal of the CNSL.

Traditionally, the kernel is removed for the CNS manually. However, to improve the deshelling process, several methods have been adopted [1,8,10], which include, among others, soaking the nuts in water to improve the moisture content thereby reducing the scorching and cracking tendencies during roasting. Roasting the nuts makes the shell brittle and loosens the kernel from the shell easily. Also, CNSL is released during the roasting [10]. Figure 5 shows the nut-processing stages.

1.3. Extraction of CNSL

Several methods are available in the literature for the extraction of CNSL. Oil extraction efficiency varies with the method adopted. Raw CNS has been reported to contain over 20% oil. The oil bath process leaves about 10% of the oil as a by-product in the spent shell. However, using an expeller extraction process, further quantities of oil may be recovered from the spent CNS.

The extraction processes can be classified into two basic types: those that involve heating and those that are done in cold or room temperature. The heating process (roasting) can be achieved by open recipients or drums [11]. In a thermomechanic (hot oil) process, the cashews can be heated by the actual CNSL [12]. In the cold process, the CNSL can be obtained by extrusion in solvents or by pressing. The cashew's liquid so obtained is denoted as natural CNSL and that extracted by the hot method is called technical CNSL [13].

Various hot methods reported in the literature for the extraction of CNSL from cashew nut shells (CNS) include, open pan roasting, drum roasting and hot oil roasting, while cold extrusion, solvent extraction, etc. are cold methods. References [14] and [15] reported CNSL extraction through pyrolysis. The extraction of CNSL using supercritical carbon dioxide has also been reported by [16,17]. Solute extracted using supercritical carbon dioxide in the extraction process is generally significantly different from those obtained by other conventional alternative processes.

Extraction methods affect the selectivity towards various compounds in CNS [18]. High concentrations of mono-unsaturated cardanol were extracted with Supercritical water (SCW) and soxhlet extraction methods (See Table 1).

Extraction fluid, substrate and operating conditions affect the extract composition (Table 2). CNSL directly obtained from CNS by SC extraction contains traces of acids whereas the same obtained from CNSL through heat exchanger mainly contains cardanol along with acids and alkanes. The oil composition of CNSL obtained through pyrolysis of CNS includes phenols and substituted phenols along with cardanol. For resin production, the CNSL obtained through pyrolysis route is the most suitable feed for extraction by SC method [3].

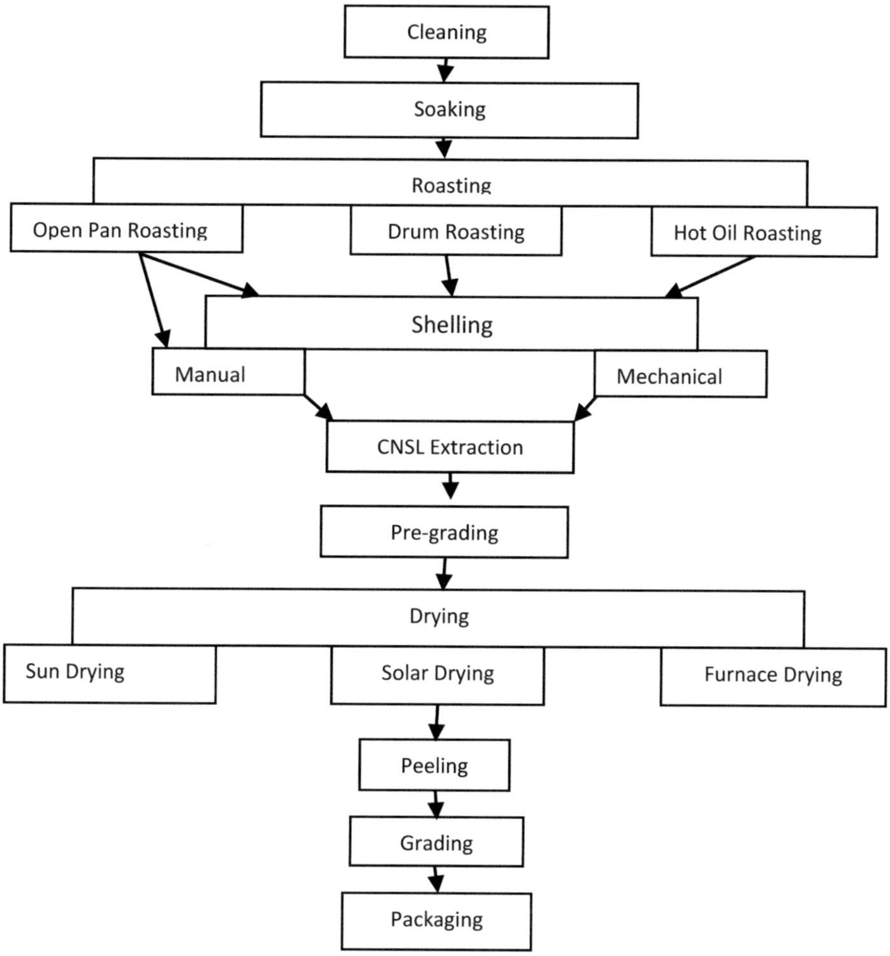

Figure 5. Flow diagram of cashew nut shell processing

Constituents	Soxhlet extraction		SCW extraction	$SC\text{-}CO_2$ extraction	Two-step extraction	
	Hexane	Methanol			Hexane-SCW	Methanol-SCW
Monounsaturated cardanol	26.64	27.19	68.93	1.09	48.84	49.22
Saturated cardanol	1.22	1.45	2.4	1.52	1.04	1.84
Unknown hydrocarbon	–	–	3.42	–	1.57	1.42
Unknown hydrocarbon	–	–	3.29	–	1.51	1.37
Octacosene	–	–	4.91	–	2.25	2.04
Stigmasterol	–	–	3.44	–	1.57	1.43

Constituents	Soxhlet extraction		SCW extraction	$SC-CO_2$ extraction	Two-step extraction	
	Hexane	Methanol			Hexane-SCW	Methanol-SCW
Monounsaturated anacardic acid	27.90	19.06	9.47	52.24	18.64	11.4
β-Sitosterol	1.38	1.86	1.63	1.63	0.75	1.76
Monounsaturated cardol	1.05	–	–	–	0.57	–
Saturated cardol	20.31	34.88	1.85	29.58	9.76	12.64
Di-unsaturated cardol	7.62	10.39	–	6.72	4.13	6.07
Unknown hydrocarbon	5.7	–	–	–	–	3.33
Triacontene	3.33	3.41	–	4.31	1.85	5.45
Others	4.85	1.76	0.66	2.91	7.52	2.03

Source: [18]

Table 1. Compound distribution of CNSL obtained by various extraction methods.

Operating Parameter (P bar/T°C)	SC-CNSL from CNS	SC-CNSL from CNS Obtained through Heat Exchanger Unit	SC-CNSL from Pyrolysis CNSL
200/60	Hexadecanoic acid (0.71) Oleic acid (0.62) Cardanol-C13 (0.69) Cardanol-C15 (84.20) Methyl Cardanol (2.83)	Penthadecane (0.67) Hexadecane (0.67) Heptadecane (0.77) 2,6,10,14 tetra methyl pentadecane (0.65) 8 methyl heptadecane (0.70) Hexadecanoic acid (0.61) Elicosane (0.47) Cardanol (62.31) Cardanol diene (31.24)	3-ethyl phenol (7.64) 2-methyl benzaldehyde (20.51) 3-butyl phenol (1.39) Cardanol diene (48.61)
225/60	Hexadecanoic acid (0.56) Oleic acid (0.684 Cardanol-C13 (0.65) Cardanol-C15 (86.26) Methyl Cardanol (3.11)		3-ethyl phenol (3.31) Azulene (1.89) Ethyloxybenzene (11.21) Acenaphthylene (7.12) 3-butyl phenol (2.28) Cardanol diene (66.82)
250/60	Hexadecanoic acid (0.58) Cardanol-C13 (7.42) Cardanol-C15 (64.89)	Penthadecane (0.04) Heptadecane (0.08) tetradecanal (1.99)	4-ethyl phenol (5.01) Propyl benzene (11.96) Acenaphthylene (2.83)

Operating Parameter (P bar/T°C)	SC-CNSL from CNS	SC-CNSL from CNS Obtained through Heat Exchanger Unit	SC-CNSL from Pyrolysis CNSL
	Methyl Cardanol (3.03) Diethyl Phthalate (12.34)	cardanol 13 (1.62) Hexadecanoic acid (1.27) Cardanol (81.54) 2 methyl Cardanol diene (2.97)	3-pentyl phenol (3.18) 3-butyl phenol (1.35) Cardanol (64.91)
300/60	Hexadecanoic acid (0.80) Oleic acid (0.67) Cardanol-C13 (2.32) Cardanol-C15 (61.34) Cardanol c17 (2.06) Diethyl Phthalate (13.35)	Hexadecanoic acid (1.01) Oleic acid (0.98) Cardanol diene (81.94) 2 methyl Cardanol diene (2.97)	2-ethyl phenol (4.89) 2-methyl benzaldehyde (8.73) Cardanol (80.50)

Source: [3]

Table 2. Chemical characterization of CNSL extracted by SCFE

Purification or refining of CNSL involves vacuum distillation of raw CNSL. However, supercritical fluid extraction SFE of CNSL can offer selective separation of the components present in CNSL by varying the operating parameters [19].

2. Experimental materials and methods

2.1. Sample collection and pretreatment

Cashew was obtained from local markets in Ile-Ife, Osun State and a commercial farm in Owu-Ijebu in Ogun state. The shell was detached from the apple, washed and sun-dried prior to deshelling the nut. The cashew nut shell was further oven-dried and milled using a ball mill at the Petrochemical laboratory of the Department of Chemical Engineering, Obafemi Awolowo University, Ile-Ife. The milled CNS was sieved to obtain a particle size range of 5 mm. The ground sample was stored at 4°C in a refrigerator for further experimentation.

2.2. Extraction of the CNS oil

The pyrolysis of the cashew nut shells was carried out in the pyrolysis reactor designed and fabricated at the Chemical Engineering Department of Obafemi Awolowo University, Ile-Ife. About 150 g of the ground and dried CNS sample was placed in the reactor and heated. The reactor is connected to a condensing unit which has several condensers in series, in order to effectively recover the exhaust gases. The experimental rig is as shown in figure 6.

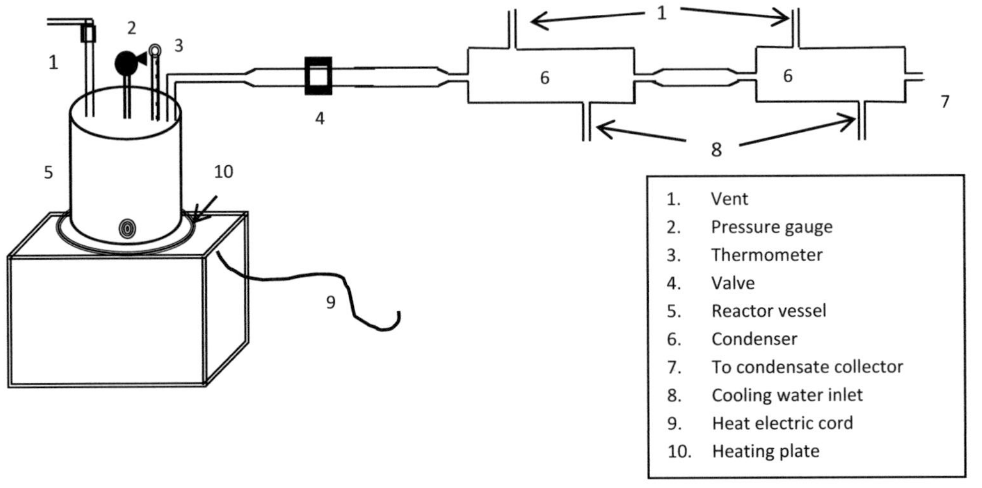

Figure 6. The experimental rig set-up.

The temperature of the reactor tank and that of the exhaust gas stream were measured and recorded at regular intervals. The volatiles which condensed at the condensing Unit were collected in several glass bottles. The heating was stopped when no significant changes in temperature and gas condensates was observed. The reactor was then allowed to cool for about 6 h; the ash and char were separated, weighed and recorded. The mass of pyrolysis oil produced was determined by mass balance calculation of the whole recovery system, before and after the experiment. For the purpose of comparison, 10 g of the CNS sample was extracted by soxhlet extraction using *n*-hexane as a solvent.

2.3. Characterization of the extracted CNSL samples

Moisture content was determined by standard Dean and Stark apparatus as outlined by the Association of Official Analytical Chemists [20]. Viscosity was determined using a capillary viscometer in water bath maintained at 40°C. A pH meter model PYE 290 was calibrated and used to determine the pH of the samples. Specific Gravity was measured using a pycnometer at 25°C. Iodine value was obtained by the method of [21].

An oxygen bomb calorimeter (model 6400) was used to determine the calorific value of the CNSL samples under adiabatic condition. The heat of the system was regulated by using an electrolyte. The calorific values of CNS and its spent cake were measured by placing them in a capsule, which was then put in the calorimeter crucible. The calorific value is given by equation 1.

$$\text{calorific value}\ \left(\text{kcal/kg}\right) = \frac{\left(W + w\right)\ \left(T_1 - T_2\right)}{X} \tag{1}$$

http://dx.doi.org/10.5772/61096

where

W = weight of water in calorimeter (kg),

w = water equivalent of apparatus,

T_1 = initial temperature of water (^{0}C),

T_2 = final temperature of water (^{0}C),

X = weight of fuel sample taken (kg)

3. Results and discussion

Table 3 shows the physicochemical characteristics of CNSL obtained including the decarboxylated CNSL. The quantity and characteristics of the CNSL extract varies with the extraction method.

Parameter	CNSL (pyrolysis)	CNSL (hexane extracted)	CNSL (decarboxylated)
Colour	Dark brown	Dark brown	Dark brown
Moisture content	4.1	6.4	3.2
Refractive index	1.693	1.688	1.698
Specific gravity	0.943	0.926	0.928
Viscosity (centipoise)	57	40	38
Ash (%)	1.22	1.53	1.50
Saponification value (mgKOH/g)	58.5	47.2	56.3
Iodine value (mg/100g)	212	236	227
Acid value (mgKOH/g)	12.4	15.6	12.9
Free fatty acid (mgKOH/g)	6.5	7.9	8.2
pH	4.8	4.5	5.2
Calorific value (kJ/g)	47.62		

Table 3. Physico-chemical characteristics of CNSL obtained including the decarboxylated CNSL.

CNSL obtained by pyrolysis although a little less in quantity than the soxhlet extraction, is denser (943 kg/m^3) with less moisture content (4.1%) and higher saponification value (58.5%). Solvent-extracted CNSL presented higher contents of cardol and anacardic acid and a lower percentage of cardanol than thermally extracted CNSL (Table 4). The higher content of cardanol, viscosity and impurity were the significant differences in CNSL obtained through pyrolysis. This gave credence to the report of Rodrigues *et al.* [22].

Components	CNSL (pyrolysis)	CNSL (hexane extracted)	CNSL (decarboxylated)
Cardol (%)	27	30	31
Cardanol (%)	54	20	65
Anacardic acid (%)	5	43	2
Others (%)	14	7	2

Table 4. Composition of the extracted CNSL

Decarboxylation of raw CNSL involves loss of a carbon dioxide molecule from anacardic acid. The anacardic acid in the oil was converted to cardanol. This is reflected in the composition variation of the components reported in table 4. It is also significant that the pyrolysis of the CNS results in loss of the carboxylic acid content of the cashew nut shell to the process. The decarboxylated anacardic acid is termed CNSL biodiesel. The biodiesel obtained from CNSL does not require further processing like Trans esterification.

CNSL biodiesel is considered as offering many advantages, including sustainability, decrease of HC, CO, NOx gas emissions and many harmful pollutants [23]. Reference [24] explains the pyrolysis procedure of CNSL. The decarboxylation of CNSL is responsible for the trend in the pH obtained for the extracts.

Most of the characteristics evaluated gave similar results as those already reported in the literature. In the present findings, the highest occurrence of cardanol was in the decarboxylated CNSL while anacardic acid and cardol occur most in the hexane-extracted and pyrolysis CNSL, respectively. This gave credence to the report of [25].

Heating CNSL decomposed the anacardic acid into cardanol and CO_2. Decarboxylation of CNSL to convert anacardic acid into cardanol could be done by heating, with an optimum heating temperature of 140°C for 1 h.

4. CNSL as petrochemical feedstock

As a substituted phenol which can take part in a variety of reactions, naturally occurring cashew nut shell liquid (CNSL), a renewable product, offers advantages over synthetics. Its innumerable application stems from the phenolic nature of its constituents, with enshrined features for transformation into high value specialty chemicals [26].

The industrial applications of CNSL-based products are numerous and include fungicide, pesticide, insecticide, brake linings, paints and primers, foundry chemicals, lacquers, cements, specialty coatings and resin. The application of CNSL component in bacteriostatic antibiotics is recently gaining attention. Their effect on plant growth, acid activity, wood preservative and pressure treatment activity are being explored. From the literature [27 - 29], much of the biological activity of CNSL has been attributed to anacardic acid: a major constituent of CNS.

Conversely, technical CNSL, as obtained by roasting or oil bath method contained mainly cardols and cardanols [27].

CNSL reduces the extent of the electrochemical processes taking place on carbon steel surface undergoing corrosion. The inhibition tendency is excellent for both static and dynamic conditions. CNSL and its derivatives have antioxidative characteristics [28 - 30]. This antioxidative tendency is related to the electron-donating nature and steric effect of the substituents. The electron-donating effect usually enhances the electron density at the phenolic oxygen of the cashew nut oil, resulting in the trapping of the radicals which normally propagates oxidation reaction. CNSL is a mixture of hindered phenols with long alkyl substitution at the *meta* position. The unsaturation on the long side chain substituents is an important factor enhancing higher anti-oxidant activity of CNSL. CNSL films are thermally stable. This stability increases in the presence of thiophosphate ester additives.

Olefin metathesis (OM) reaction on cardanol is an important class of reactions that allows for the synthesis of new olefins that are sometimes impossible to prepare via other methods [31]. Cardanol- a renewable, low cost, main constituent (about 85%) of CNSL possesses characteristic long alkyl chains in *meta* position of its phenol ring [12] that promotes this OM reaction, and also influences several chemical transformations leading to novel functionalities [31].

Cardanol based polyhydrins improve the photo-catalytic activity of bare TiO_2. The porphyrins are brown-red sticky solids, very soluble in $CHCl_3$ or CH_2Cl_2 [32]. This substituted cardanol-based porphyrin and their metallic complexes have been used extensively in photo degradation of toxic, bio-refractory 4-nitro phenol in water, which is dangerous for the ecosystem and for human health [32]. CNSL is also useful in fuel blends and fuel mixtures [33] and for producing diesel oil [34]. Cracking of CNSL with a molecular sieve at 500° C for 2 h generates brown coloured liquid product. This is diesel fuel.

Cashew Nut Shell Liquid (CNSL) was used as an alternative fuel for diesel engine. The viscosity of CNSL is 30 - 35 times higher than diesel; hence different blends of CNSL would have different properties and application. Also, modification of the oil and its application condition such as, injection pressure, injection timing and preheating the oil also optimizes the performance of the engine. There were indications that preheating of CNSL25 blends at 200 kg/cm^2 injection pressure and 28° injection timing gives encouraging results suitable for commercial purposes [35]. CNSL as a bioadditive in engines increases the durability of the equipment. Hence, the application of CNSL as a bioadditive will reduce the dependency on petroleum products besides preserving the environment by lowering pollutant residues from fuel combustion products. Some of the properties of diesel, biofuel and ethanol are presented against CNSL in Table 5.

CNSL has higher density than diesel. It can be reduced by degumming and trans esterification. Cetane for CNSL is expected to be poor due to the presence of aromatic compounds. The usual C:H:O ratio for vegetable oils is 78:12:10 whereas for CNSL it is 80:12:8; hence, it justifies the higher calorific value of CNSL (47 MJ/kg), whereas for diesel it is 42 MJ/kg. Also, ash content is well within the limits for CNSL. The water content of CNSL is quite high. The sulphur content does not exceed 0.006% for any vegetable oil. Thus it is anticipated that CNSL has no sulphur content. The flash point for CNSL is 164°C; this is higher than the flash point for diesel and represents the higher starting ignition temperatures and compression of CNSL [35].

Properties	Diesel*	Ethanol*	CNSL
Density (kg/m^3)	0.84	0.789	0.9326
Kinematic viscosity (cost)	2-5	1.19	17.2
Calorific value (kJ/kg)	42000	30000	47600
Flash point (°C)	62	16	193
Auto-ignition temperature (°C)	210	362	206

*Source: [28]

Table 5. Fuel properties

The physicochemical properties of cashew nut shell liquid (CNSL) make it potentially useful for improved adhesion between bitumen and aggregates. Thus, the asphalt resulting from bitumen modified with CNSL has better stripping resistance and satisfactory mechanical properties compared with conventional asphalts. The CNSL, as an additive, prevents stripping in the asphalts, thereby contributing to its moisture damage resistance [36].

CNSL has also been used in production of asbestos with free friction composition for brake linings [37]. In combination with cardanol, it is used extensively in automotive break-lining applications [38]. CNSL/cardol-based surface coatings possess excellent gloss and surface finish with a high level of toughness and elasticity.

Valuable chemicals result from the pyrolysis of CNSL at high temperatures (450°C to 750°C) and at short residence times [39]. The products of pyrolysis include tar, phenols, hydrocarbons and gases with high petrochemical values.

On the other hand, cardanol which is highly resistant to softening action of mineral oil, acids, alkalis, microbes, termites and insects is widely used in the coating and resin industry. Reference [40] reported that technical CNSL may also contain phytosterol, such as stigma sterol and β-sitosterol, which have cholesterol-lowering properties.

Cardanol polysulfide (CPS) has been used as a vulcanizing agent for natural rubber as reported by [41]. Coordination bond linkage in cardanol-aldehyde condensation polymers improved its physicomechanical properties. This positively influences the anticorrosive properties and stability of the polymeric compound at high temperature [42]. The blend of cardanol-based novolac-type phenolic resins, with commercial epoxy or isocynate monomers produces thermoset polymers [43]. Also, in [44], the conversion of CNSL-based cardols (6-alkenylresorcinols) into lasiodiplodin, a naturally occurring 12-membered orsellinic acid type macrolide, exhibits plant growth-regulating and antileukemic properties.

Immature CNSL (iCNSL) has excellent protective activities in strains of *S. cerevisiae* against oxidative damage induced by hydrogen peroxide and inhibits acetyl cholinesterase activity [45]. The high in vitro antibacterial activity of cashew nut shell liquid (CNSL) has been attributed to its anacardic acid content [46]. Both anacardic acid and cardol were reported to have antitumor [47-49], antimicrobial [48], urease inhibitory [50] and lipoxygenase activities [51]. Gram positive bacteria, which cause tooth decay, acne, tuberculosis, *Streptococcus pneumonia, Francisella tularensis,* and leprosy are killed by anacardic chemicals [52].

Anacardic acid has also been found to reduce the expression of survivin and X-linked inhibitor of apoptosis protein, anti apoptotic proteins associated with cellular survival and radio resistance, and radio-sensitized pituitary adenoma cells [53]. Also, these acids have been used effectively against tooth abscesses due to their lethality to bacteria.

The monophenolics of CNSL have been found to form stable quaternary nitrogen compounds which are soluble in water, odourless and act as surface-active agents [54]. These properties including their high bacterial activity, have made them significantly useful as germicides, disinfectants and sanitizing agents in food and dairy industries. Also, sodium salts of anacardic acid, for example disodium anarcadates, are anionic surfactant and could be used as bactericidal surfactant while sodium anarcadate is useful for the control of vector mosquitoes which are responsible for causing dengue fever [55].

Energy studies established that cashew nut by-products CNSL, its isolates and spent CN Shell compare quite favourably with conventional fuels in terms of energy content. CNSL, cardanol and cardol, which are liquid by-products, were superior to conventional liquid and solid fuels while spent CN shell was more superior to a number of liquid fuels, e.g. ethanol and methanol, as well as fire wood in terms of energy content. This material is advocated for use by cashew nut processors and other industries as fuel to reduce environmental pollution. This material is not only a cleaner fuel but also has low ash and is renewable [56].

Substitution of phenol with CNSL in both resol and novolacs decreased the tensile strength, but improved the impact strength of wood flour-based laminate of these resins. Thermal properties of phenol-cardanol-based resin depletes with increasing cardanol content in the resin [57, 58].

The synthesis of polymers from renewable resources has attracted considerable attention worldwide due to its potential attribute as a substitute for petrochemical feedstock. CNSL may act as a potential raw material for the manufacture of polymers in the 21st century [58] due to its components and depletion in petroleum production which increasingly faces exhaustion. Long-chain, *m*-substituted phenol in CNSL is highly reactive. Thus a wide variety of resins are synthesized from CNSL, such as polyesters, phenolic resins, epoxy resins, polyurethanes, acrylics, vinyls and alkyds to mention a few [59].

CNSL contains four major components: 3-pentadecenyl phenol (cardanol), 5-pentadecenyl resorcinol (cardol), 6-pentadecenyl salicylic acid (anacardic acid) and 2-methyl 5-pentadecenyl resorcinol (2-methyl cardol), which can be good replacement for synthetic resins, owing to the current climate of diminishing petroleum reserves. Cardanol, which is a phenol derivative, has a meta substitute of a C15 unsaturated hydrocarbon chain with one to three double bonds [31, 32, 60, 61]. Being a major component of CNSL, its abundant production promised renewable feedstock for the petrochemical industry with a total production of CNS approaching 2.6 million tons annually [1].

Resoles and novolacs resins are the major products of reaction of CNSL with formaldehyde. While novolac results from acid-catalyzed reactions, resoles are a product of base-catalyzed reactions.

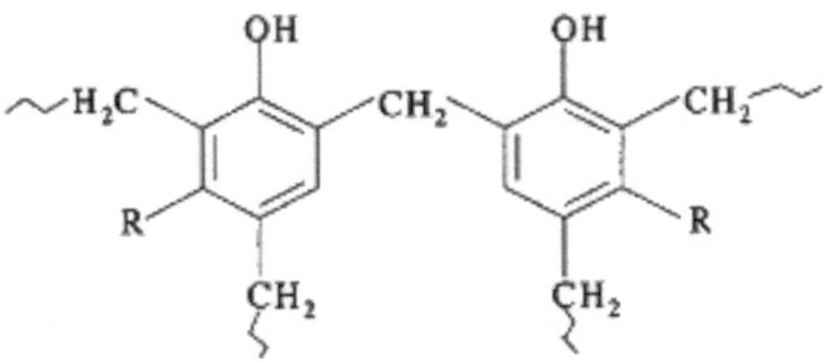

CNSL can be polymerized by different methods. This includes addition polymerization through the side-chain double-bonds using cationic initiators such as diethylsulphate, condensation polymerization through the phenolic rings with aldehydic compound, e.g. formaldehyde, oxidative polymerization, polymerization after chemical modification to produce specialty properties, etc. The most common method is condensation reaction with formaldehyde. The CNSL-based polymers offer advantages such as improved flexibility and reduced brittleness, solubility in organic solvents, improved process ability, compatibility with other polymers, high performance and resistance to microbes, insects and termites.

CNSL were successfully applied as a thermosetting wood adhesive for wood panels to reduce formaldehyde emission. These renewable phenolic compounds give excellent adhesive performance, good moisture resistance and tend to give a lower formaldehyde emission than UF resin [62, 63].

4.1. Environmental impact

CNSL may be a weak promoter of carcinogenesis but no mutagenic or carcinogenic activity has been reported [64]. Epidemiological studies suggested that CNSL may contribute to oral sub-mucous fibrosis [65]. In addition, its phenolic components exerted several biological activities, including antioxidative properties [66, 67], inhibition of acetyl cholinesterase and membrane perturbation [68, 69].

There is no direct evidence regarding the toxicity of CNSL or its major phenolic components. However, the effluent generated during the processing of the cashew nut could be considered potentially harmful to the environment due to its high phenol content. The hazardous effects of phenolic compounds have been extensively studied. According to [70], a concentration above 1 mg/L can affect aquatic life, while [71] recognizes deleterious effects at concentrations as low as 1 μg/mL.

As previously mentioned, data on the chemical composition of the cashew nut industry effluent are scarce, but preliminary analysis showed high phenol content due to the cardol, cardanol and anacardic acid of CNSL. The high toxicity observed for the isolated phenols, cardol and cardanol potentially contribute to the toxicity of the cashew nut industry effluent [72].

5. Conclusion

The report has highlighted various chemicals obtainable from CNSL both directly and indirectly. This inexhaustible listing presents CNSL as a very important, reliable source of raw material for petrochemical industry. It is a good, promising supplement and/or alternate to petroleum, which is currently facing depletion globally.

Varied CNSL compositions with varying modes of extraction are a vantage opportunity with potential for multiple applications.

Though CNSL contains phenolics, its products are highly environmental friendly. However, efficient treatment strategy may be inevitable to reduce environmental impact associated with the production industry.

Acknowledgements

The author is grateful to the laboratory and technical staff of the Department of Chemical Engineering, Obafemi Awolowo University, Ile-Ife, Nigeria, for the support rendered in carrying out the work.

Author details

Elijah A. Taiwo*

Address all correspondence to: etaiwo@oauife.edu.ng,eataiwo@yahoo.com

Department of Chemical Engineering, Obafemi Awolowo University, Ile-Ife, Osun State, Nigeria

References

[1] Subbarao Ch N V, Krishna Prasad K M M, Prasad V S R K. Review on applications, extraction, isolation and analysis of cashew nut shell liquid (CNSL). The Pharma Research Journal. 2011; 6(1): 21- 41. DOI:10.1007/s11095-015-1664-8

[2] Olife I C, Jolaoso M A, Onwualu A P. Cashew processing for econonic development in Nigeria. Agricultural Journal. 2013; 8(1), 45 - 50. DOI: 10.3923/aj.2013.45.50.

[3] Patela R.N, Bandyopadhyayb S, Ganesh A. Selective extraction of cardanol and phenols from cashew nut shell liquid obtained through pyrolysis of cashew nut shells. In: Proceedings of the Indian Chemical Engineering Congress, Novel Separation Processes Session (CHEMCON-2005) 14 -17 December 2005; New Delhi, India.

[4] Akinhanmi T F, Atasie V N, Akintokun P O. Chemical composition and physicochemical properties of cashew nut (Anacardium occidentale) oil and cashew nut shell liquid. Journal of Agriculture, Food and Environmental Sciences. 2008; 2(1), 10pp

[5] Pushparaj T, Ramabalan S. Influence of CNSL biodiesel with ethanol additive on diesel engine performance and exhaust emission. International Journal of Mechanical Engineering and Technology (IJMET). 2012; 3(2), 665 - 674

[6] Hammed L A, Antique J C, Adedeji A.R. Cashew nuts and production development in Nigeria. American-Eurasian Journal of Scientific Research. 2008; 3(1): 54 - 61

[7] FAOSTAT data, 2013 (last accessed by www.top5of anything.com: December 2014).

[8] Ojewola G S, Okoye F C, Agbakuru I. Performance value of cashew nut meal for soya bean meal in finishing broiler chickens. International Journal of Poultry Science. 2004; 3(8): 513-516. DOI: 10.3923/ijps.2004.513.516.

[9] Kozubek A, Tyman J H P. Resorcinolic iipids, the natural non-isoprenoid phenolic amphiphiles and their biological activity. Chemical Reviews. 1999, 99(1):1 - 25. DOI: 10.1021/cr970464o

[10] Sue Azam-Ali. Cashew nut processing–An overview of delicious cashew nut processing and various choices in methods. Technical brief, Practical Action. Schlumacher centre, bourton on Dunsmore, Rugby, Warwickshire, U.K. 2001, 9 pages.

[11] Patel R N, Bandyopadhyay S, Ganesh A. Extraction of cashew *(Anacardium occidentale)* nutshell liquid using supercritical carbon dioxide. Bioresource Technology. 2006; 97: 847–853

[12] Mazzetto S E, Lemonaco D, Mele G. Cashew nut oil: opportunities and challenges in the context of sustainable industrial development. Química Nova. 2009; 32(3): 732-741. DOI:10.1590/S0100-40422009000300017

[13] Rodrigues F H A, França F C F, Souza J R R, Ricardo N M P S, Feitosa J P A. (2011) Comparison Between Physico-chemical properties of the technical cashew nut shell liquid (CNSL) and those natural extracted from solvent and pressing. Polímeros. 2011; 21(2): 156 -160. DOI: 10.1590/S0104-14282011005000028.

[14] Das P, Sreelatha T, Ganesh A. Bio-oil from pyrolysis of cashew nut shell characterization and related properties. Biomass and Bioenergy. 2004; 27: 265 - 275. DOI: 10.1016/S0961-9534(02)00182-4

[15] Tsamba A J, Yang W, Blasiak W. Pyrolysis characteristics and global kinetics of coconut and cashew nut shells. Fuel Processing Technology. 2006, 87: 523 - 530. DOI: 10.1016/j.fuproc.2005.12.002

[16] Shobha S V, Ravindranath B. Supercritical carbon dioxide and solvent extraction of the phenolic lipids of cashew nut *(Anacardium occidentale)* shells. Journal of Agricultural and Food Chemistry. 1991; 39: 2214 - 2217. DOI: 10.1021/jf00012a022

[17] Smith R.L, Malaluan R.M, Setianto W.B, Inomata H, Arai K. Bio based manufacture of alkyl phenols and polysaccharides from cashew nut with supercritical carbon dioxide and water. Bioresource Technology. 2003; 88: 1 - 3.

[18] Yuliana M, Ngoc Yen Tran-Thi, Yi-Hsu Ju. Effect of extraction methods on characteristic and composition of Indonesian cashew nut shell liquid. Industrial Crops and Products. 2012; 35(1): 230 – 236. DOI: 10.1016/j.indcrop.2011.07.007

[19] Patel R N, Bandyopadhyay S, Ganesh A. Economic appraisal of extraction of refined cashew nut shell liquid through supercritical fluid extraction. Journal of chromatography A. 2006; 1124: 130 -136.

[20] AOAC. Official Methods of Analysis of the Association of Official Analytical Chemists, Edited by Keneth Helrich, 15th ed. 1990.

[21] Strong F M, Koch G H. Biochemistry laboratory manual. 2nd ed. Dubuque, IA: W.M.C. Publishers. 1974.

[22] Rodrigues F H A, Feitosa J P A, Nágila M P S, Ricardo F C F, de França A, José O B, Cario M. Antioxidant activity of cashew nut shell liquid (CNSL) derivatives on the thermal oxidation of synthetic cis-1,4-polyisoprene. Journal of the Brazilian Chemical Society. 2006; 17(2): 265 -271. DOI: 10.1590/S103-50532006000200008

[23] Raghavendra Prasada S A. A Review on CNSL biodiesel as an alternative fuel for diesel engine. International Journal of Science and Research (IJSR). 2014; 3(7): 2028 -2038.

[24] Risfaheri T T I, Nur M A, and Sailah I. Isolation of cardanol from cashew nut shell liquid using the vacuum distillation method. Indonesian Journal of Agriculture. 2009; 2(1): 11 -20.

[25] Gandhi T, Patel M, Dholakiya B.K. Studies on effect of various solvents on extraction of cashew nut shell liquid (CNSL) and Isolation of major phenolic constituents from extracted CNSL. Journal of Natural Product and Plant Resources. 2012; 2 (1): 135 -142.

[26] Buchweishaija J, Mkayula L L. The effect of rotation and temperature on the inhibition performance of cashew (Anacardium occidentale L) nut shell liquid on corrosion

of carbon steel. Transactions of the Journal Engineering and Technology. 2008; 2(2): 121 -125.

[27] Sanger S H, Mohod A G, Khandetode Y P, Shrirame H Y, Deshmukh A S. Study of carbonization for cashew nut shell. Research Journal of Chemical Sciences. 2011; 1(2): 43 -55

[28] Velmurugan A, loganathan M. Effect of ethanol addition with cashew nut shell liquid on engine combustion and exhaust emission in a DI diesel engine. International Journal of Engineering Science and Technology. 2012; 4(7): 3316 -3328.

[29] Oliveira M S, Morais S M, Magalhães D V, Batista W P, Vieira I G, Craveiro A A, de Manezes J E, Carvalho A F, de Lima G P. Antioxidant, larvicidal and anti-acetyl cholinesterase activities of cashew nut shell liquid. Acta Tropica. 2011; 117(3): 165-170. DOI: 10.1016/j.actatropica.2010.08.003

[30] Lomonaco D, Cangane F Y, Mazzetto S E. Thiophosphate esters of cashew nutshell liquid derivatives as new antioxidants for poly(methyl) methacrylate. Journal of Thermal Analysis and Calorimetry. 2011; 104(3): 1177 -1183.

[31] Vasapollo G, Mele G, Sole R D. Cardanol based material as natural precursors for olefin metathesis. Molecules.2011; 16: 6871 -6882. DOI: 10.3390/molecules16086871

[32] Vasapollo G, Mele G, Sole R D, Pio I, Li J, Mazzetto S E. Use of novel cardanol-porphyrin hybrids and their TiO_2-based composites for the photodegradation of 4-nitrophenol in water. Molecules. 2011; 16: 5769 -5784. DOI:10.3390/molecules16075769

[33] Fernando José Araújo da Silva and José Everardo Xavier de Matos. A note on the potential of CNSL in fuel blends for engines in Brazil. Revista Tecnologia. 2009; 30 (1): 89 -96.

[34] Paisal Nakpipat and Hiroo Niiyama Diesel oil from cashew nut shell liquid. Regional Symposium on Chemical Engineering. 1999.

[35] Solanki J H, Javiya T V. Cashew nut shell liquid fuel a substitute for diesel fuel to be used in C.I. engine. International Journal of Advanced Research in Science, Engineering and Technology. 2012; 1 (2): 8 -12

[36] Ribeiro E A, Fernando Felipe Lopes Antunes, Verônica Teixeira Franco Castelo Branco, Sandra Aguiar Soares, Jorge Barbosa Soares. Evaluation of moisture damage in asphalt containing cashew nut Shell liquid (CNSL) modified bitumen. In: Proceedings of the 5th Euro asphalt & Euro bitumen Congress; 13 -15th June 2012, Istanbul. 10pages

[37] Ganguly A, Raji G. Asbestos free friction composition for brake linings. Bulletin of Materials Science. 2008; 31(1), 19 -22. DOI: 10.1007/s12034-008-0004-6

[38] Murthy B.G.K, Sivasamban M.A. Recent trends in CNSL utilization. Cashew Research and Development: Proceedings of the International Cashew Symposium, Cochin, Kerala, India, 1985, 12 -15.

[39] de Lasa H I, Afara S. Processing of cashew nut shell liquid. Final Technical Report; Mozambique. Chemical Reactor Engineering Centre, Faculty of Engineering Science, University of Western Ontario, London. August 1, 1995.

[40] Andrade T J A S, Araújo B Q, Citó A M G L, Silva J, Saffi J, Richter M F, Fer-raz A B F. Antioxidant properties and chemical composition of technical cashew nut shell liquid (tCNSL). Food Chemistry. 2011; 126(3): 1044–1048. DOI: 10.1016/j.foodchem. 2010.11.122

[41] Leerawan K, Narumon S, Nanthita K, Jumreang T, Sophon R. Cardanol polysulfide as a vulcanizing agent for natural rubber. Journal of Scientific Research, Chulalongkorn University. 2005; 30(1): 23 - 40.

[42] Jinhuo L, Bingham H U. Study on the cardinal-aldehyde condensation polymer containing boron-nitrogen coordinate bond. Chinese Journal of Polymer Science. 1998; 16(3): 219 -225.

[43] D'Amico D, Martina F, Ingrosso G, Mele G, Maffezzoli A, Tarzia A, Stifani C. Development and characterization of jute reinforced natural matrix composites. International Journal of Materials and Product Technology. 2009; 36(1 - 4):155 -165. DOI: 10.1504/IJMPT.2009.027827

[44] Santos M L, Magalhães G C. Utilization of cashew nut shell liquid from *Anacardium occidentale* as starting material for organic synthesis: A novel route to lasiodiplodin from Cardols. Journal of the Brazilian Chemical Society. 1999; 10(1): 13-20. DOI: 10.1590/S0103-50531999000100003

[45] De Lima S.G, Feitosa C.M, Citó A.M.G.L, Neto J.M., Lopes J.A.D, Leite A.S, Brito M.C, Dantas S.M.M, Melo Cavalcante A.A. Effects of immature cashew nut-shell liquid (*Anacardium occidentale*) against oxidative damage in *Saccharomyces cerevisiae* and inhibition of acetyl cholinesterase activity. Genetics and Molecular Research. 2008; 7(3): 806 -808.

[46] Eichbaum F. Biological properties of anacardic acid and related compounds. Memórias do Instituto de Butantan. Sao Paulo. 1946; 19:69 -96.

[47] Kozubek A, Zamowski R, Stasiuk M, Gubernator J. Natural amphiphilic phenols as bioactive compounds. Cellular and Molecular Biology Letters. 2001; 6: 351–355. DOI: 10.2478/s11756-012-0030-0 0.70

[48] Kubo I, Ochi M, Viera P. C, Komatsu S. Antitumor agent from the cashew (*Anacardium occidentale*) apple juice. Journal of Agricultural and Food Chemistry. 1993; 41: 1012 -1015. DOI: 10.1021/jf000309035

[49] Melo Cavalcante A A, Rübensam G, Erdtmann B, Brendel M, Henriques J A P. Cashew (*Anacardium occidentale*) apple juice lowers mutagenicity of aflatoxin B_1 in S. *typhi-*

murium TA102. Genetics and Molecular Biology. 2005; 28: 328 – 333. DOI: 10.1590/S1415-47572005000200025.

[50] Kubo J, Lee J R, Kubo I. Anti-*Helicobacter pylori* agents from the cashew apple. Journal of Agricultural and Food Chemistry. 1999; 47(2): 533 –537. DOI: 10.1021/jf9808980

[51] Ha T J, Kubo I. Lipoxygenase inhibitory activity of anacardic acids. Journal of Agricultural and Food Chemistry. 2005; 53(11): 4350 – 4354. DOI: 10.1021/jf048184e

[52] Himejima M, Kubo I. Antibacterial agents from the cashew *Anacardium occidentale* (Anacardiaceae) nut shell oil. Journal of Agricultural and Food Chemistry. 1991; 39(2): 418-421. DOI: 10.1021/jf00002a039

[53] Sukumari-Ramesh S, Singh N, Jensen M A, Krishnan M. D, Vender J R. Anacardic acid induces caspase independent apoptosis and radiosensitizes pituitary adenoma cells; Laboratory investigation. Journal of Neurosurgery. 2011; 14(6) :1681 -1690.

[54] Gulati A S, Krishnamachar V.S, Subbarao B C. Supercritical carbon dioxide extraction of Indian cashew nut shell. Indian Journal of Chemistry. 1964; 2: 114 -117.

[55] Farias D F, Cavalheiro M G, Viana S M, De Lima P G, Da Rocha-Bezerra C B, Ricardo M.P.S, Carvalho F U. Insecticidal action of sodium anacardate from Brazilian cashew nut shell liquid against *Aedes aegypti.* Journal of the American Mosquito Control Association. 2009; 25(3): 386-389. DOI: 10.2987/08-5851.1

[56] Mwangi P M, Aule C, Thiong'o G T. Energy studies of some cashew nut by-products in Kenya. International Journal of Advanced Research. 2013; 1(8): 880 -887.

[57] Reghunadhan Nair C P, Bindu R.L, Joseph V C. Cyanate esters based on cardanol modified-phenol-formaldehyde resins: synthesis and thermal characteristics. Journal of Polymer Science: Part A: Polymer Chemistry. 1995; 33(4): 621 -627.DOI: 10.1002/pola.1995.080330403

[58] Bhunia H P, Nando G. B, Chaki T. K, Basak A, Lenka S, Nayak P. L. Synthesis and characterization of polymers from cashew nut shell liquid (CNSL). European Polymer Journal. 1999; 35: 1381 -1391. DOI: 10.1016/S0014-3057(98)00225-0.

[59] Balgude D and Sabnis A. S. CNSL: an environment friendly alternative for the modern coating industry. Journal of Coatings Technology and Research. 2014; 11(2): 169-18334. DOI: 10.1007/s11998-013-9521-3

[60] Kim Y H, An E. S, Park E. S, Song B K. Enzymatic epoxidation and polymerization of cardanol obtained from a renewable resource and curing of epoxide-containing polycardanol. Journal of Molecular Catalysis B: Enzymatic. 2007; 45: 39 - 44. DOI: 10.1016/j.molcatb.2006.11.004

[61] Bhunia H P, Jana R. N, Basak A., Lenka S, Nando G. B. Synthesis of polyurethane from cashew nut shell liquid (CNSL), a renewable re-source. Journal of Polymer Sci-

ence Part A: Polymer Chemistry. 1998; 36(3): 391 - 400. DOI: 10.1002/(SICI)1099-0518(199802)36:3<391:: AID-POLA3>3.0.CO;2-V

[62] Kim S. The reduction of formaldehyde and VOCs emission from wood-based flooring by green adhesive using cashew nut shell liquid (CNSL). Journal of Hazardous Materials. 2010; 182: 919 - 922. DOI:10.1016/j.jhazmat.2010.03.003

[63] Lee J H, Jeon J, Kim S. Green adhesives using tannin and cashew nut shell liquid for environment-friendly furniture materials. Journal of the Korea Furniture Society. 201; 22(3): 219 -229

[64] George J, Kuttan R. Mutagenic, carcinogenic and cocarcinogenic activity of cashew nut shell liquid. Cancer Letters. 1997; 112(1):11 -16. DOI: 10.1016/S0304-3835(96)04540-5

[65] Varghese I, Rajendran R, Sugathan C K, Vijayakumar T. Prevalence of oral sub mucous fibrosis among the cashew workers of Kerala-South India. Indian Journal of Cancer. 1986. 23: 101 - 104.

[66] Trevisan M T S, Pfundstein B, Haubner R, Wurtele G, Spiege D, Halder B, Bartsch H, Owen R W. Characterization of alkyl phenols in cashew (*Anacardium occidentale*) products and assay of their antioxidant capacity. Food and Chemical Toxicology. 2006; 44(2): 188-197. DOI: 10.1016/j.fct.2005.06.012

[67] Façanha M A, Mazzetto S A, Carioca J O, Barros G G. Evaluation of antioxidant properties of a phosphorated cardanol compound on mineral oils (NH10 and NH20). Fuel 2007; 86(15): 2416 -2421. DOI: 10.1016/j.fuel.2007.01.034.

[68] Stasiuk M, Kozubek A. Membrane perturbing properties of natural phenolic and resorcinolic lipids. FEBS Letters. 2008; 582: 3607-3613. DOI: 10.1016/j.febslet.2008.09.039

[69] Stasiuk M, Bartosiewicz D, Kozubek A. Inhibitory effect of some natural and semisynthetic phenolic lipids upon acetyl cholinesterase activity. Food Chemistry. 2008; 108: 996 -1001. DOI: 10.1016/j.foodchem.2007.12.011

[70] Veeresh G S, Kumar P, Mehrota I. Treatment of phenol and cresols in up flow anaerobic sludge blanket (UASB) process: a review. Water Research. 2005; 39(1): 154 - 170. DOI: 10.1016/j.watres.2004.07.028

[71] Newman M C, Unger M A. Fundamentals of Ecotoxicology. 2nd ed. Lewis Publishers; 2003. 433 p.

[72] Pimentel M F, De Lima D P, Martins L R, Beatriz A, Santaella S T, Costa Lotufo L V. Eco toxicological analysis of cashew nut industry effluents, specifically two of its major phenolic components, cardol and cardanol. Pan-American Journal of Aquatic Sciences. 2009; 4(3): 363 - 368

Chapter 2

Natural Gas Geochemistry in the Offshore Nile Delta, Egypt

Mohamed Abdel-Aziz Younes

Additional information is available at the end of the chapter

http://dx.doi.org/10.5772/60575

Abstract

The offshore Nile Delta basin is considered as one of the most promising province in Egypt which has an excellent potential gas and condensate reserves for future exploration. This study aimed to characterize the origin of natural gas and to determine whether it is biogenic, thermogenic or mixed processes which are considered the dominant generation pathways. Regional geochemical studies were conducted on 20 test gas samples produced from the Miocene and Plio-Pleistocene reservoirs of Abu Qir, N. Abu Qir, Temsah, Wakkar and Port Fouad fields representing the western and the eastern Nile Delta province. The geochemical analyses revealed that the main constituents of the natural gases are methane (96.2%–99.37%) with minor contribution from ethane and propane. Methane carbon isotopic composition ranges between -65.6‰ and -40.3 ‰ PDB displaying a strong indication for gas mixing of thermogenic and early microbial methane. The gas chromatography_mass spectrometry carried out on the produced natural gases exhibits high oleanane index that ranges between 19% and 42%, the medium concentration of moretane index between 11% and 16% and the absence of gammacerane index indicating that the natural gases were derived from siliciclastic source rocks containing type III kerogen of terrestrial origin and higher land plants input of Tertiary age. The calculated maturity parameters of the studied natural gas proportions based on various sterane isomerisation distributions, i.e. $C_{29}\alpha\beta\beta/(\alpha\beta\beta+\alpha\alpha\alpha)$_and $C_{29}\alpha\alpha\alpha 20S/(S+R)$ reached 0.6 and 0.5 respectively indicating a medium stage of thermal maturation equivalent to the main peak of oil generation window (0.85 Ro%). The sterane isomerization ratios may reflect the rapid rate of subsidence and sedimentation in the Nile Delta and appear to have been generated during the early stage of source rock maturation.

Keywords: biogenic, thermogenic, Nile Delta, biomarkers, Egypt

INTECH
open science | open minds

© 2015 The Author(s). Licensee InTech. This chapter is distributed under the terms of the Creative Commons Attribution License (http://creativecommons.org/licenses/by/3.0), which permits unrestricted use, distribution, and reproduction in any medium, provided the original work is properly cited.

1. Introduction

1.1. Location of the study area

The Nile Delta and the offshore Mediterranean Sea basin are hydrocarbon rich provinces that have generated natural gas and condensate from siliciclastic reservoirs ranging in age from Plio_Pleistocene to Miocene. Currently, these areas are the most active exploration and development province in Egypt. The proven huge reserves of the Pliocene gas discoveries made in the last decade firmly established the Pliocene sequence as a primary hydrocarbon potential target for the exploration activities in the Nile Delta and offshore Mediterranean region.

The study area lies in the offshore Mediterranean Sea to the north of Alexandria and Port Said cities within the northern eastern Nile Delta between longitudes 30° 10′- and 32° 54′-E and latitude 31° 25′- and 32° 00′- N. This area includes four giant natural gas fields namely; Abu Qir Field located to the north of Alexandria city and Temsah, Wakkar and Port Fouad fields located 65 kilometers to the north of Port Said in the offshore Mediterranean Sea, about 30 km from the Egyptian coast (Figure 1).

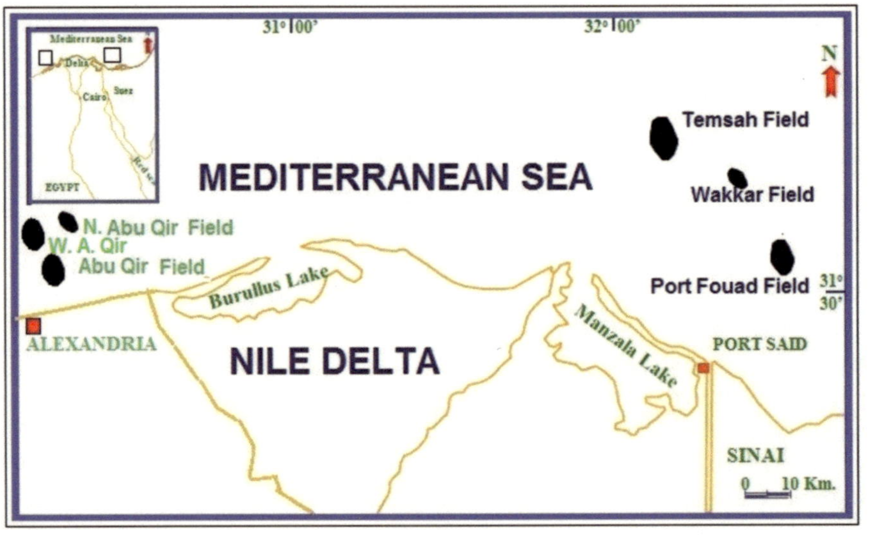

Figure 1. Location map of the study gas fields, offshore Nile Delta, Egypt.

1.2. Geologic setting and lithostratigraphy

The present Nile Delta lies on the north_external margin of the African plate. Its structural evolution is largely contemporaneous with the break–up of the African plate margin consequent to the opening of the Red Sea and the northward transition of the Arabian Peninsula [1, 2]. The northern part of the Nile Delta is characterized by a series of major tectonic features

http://dx.doi.org/10.5772/60575

with different orientations. Some of them are active during the deposition of Tertiary section while other flatten with depth, become horizontal and dies on top of the Rosetta Formation affecting only through the Kafr El-Sheikh Formation.

The Nile Delta is characterized by asymmetric fold, overthrust faults and diapers. Its dates back to the Syrian Arc system having an arcuate trend from northeast to southwest through the northern part of the Nile Delta to the Western Desert of Egypt (Figure 2).

The regional subsurface stratigraphic succession of the Nile Delta has been represented in Figure 3. This succession is discussed by many authors [3-11].

The lithostratigraphy of the Nile Delta is represented by Miocene sediments:

Moghra, Sidi Salem and Abu Madi formations (from bottom to top). In addition, the Pliocene–Pleistocene sequences are represented mainly by shale or clay with sandstone interbeds. The Miocene sequence is summarized as follows:

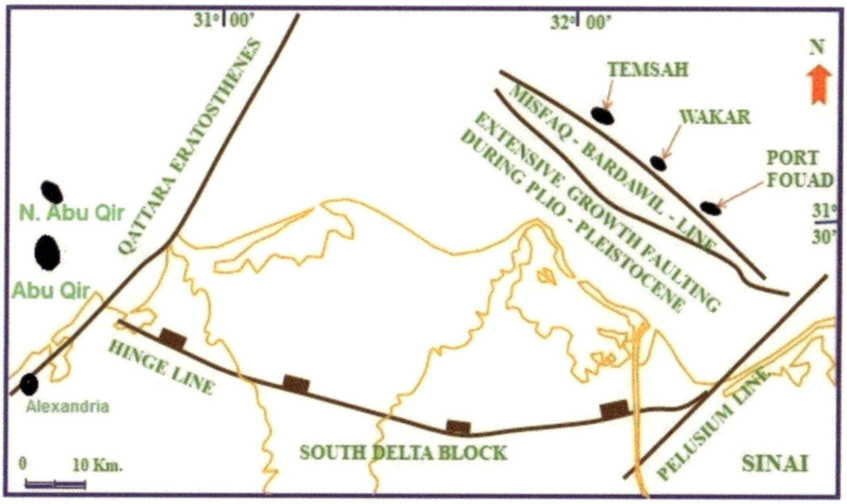

Figure 2. Regional structural setting, north Nile Delta from Mosconi et al. [38] and Hemdan et al. [39].

1.2.1. Moghra formation (Early miocene)

Moghra Formation consists of marine to fluvio–marine deposits (shale and sandstone) of presumed deltaic origin, although it is associated with marine carbonate intervals. This formation is unconformably overlain by Sidi Salem Formation.

1.2.2. Sidi Salem formation (Middle miocene)

The lower limit of this formation is not known in the central part of the Nile Delta where it overlies the Moghra Formation to the west and south. Sidi Salem Formation is composed mainly of mudstone with interbeds of marls, sandstones and siltstones. This formation

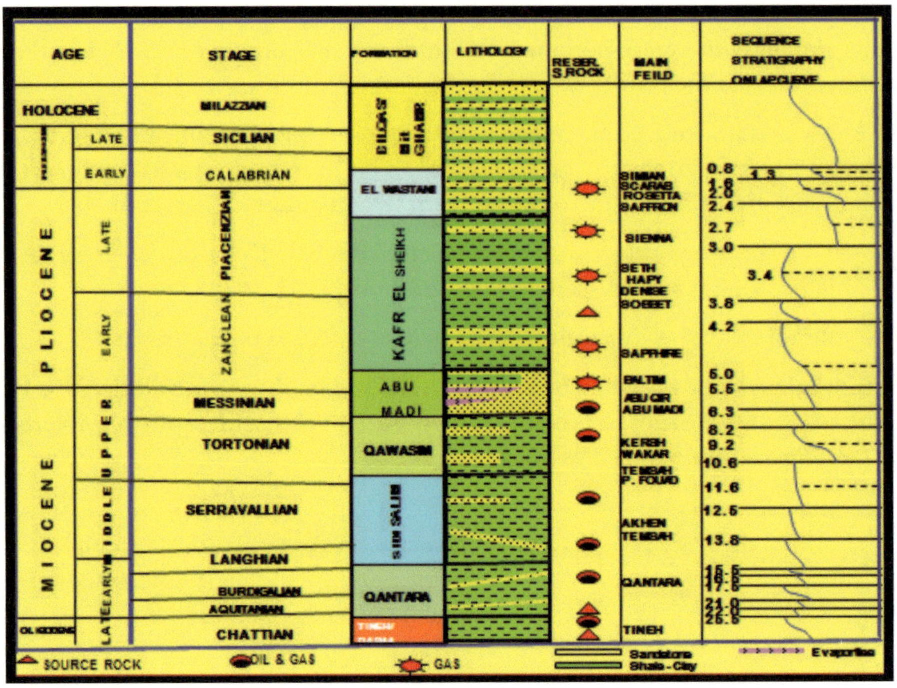

Figure 3. Lithostratigraphic succession of Northern Nile Delta, Egypt modified from Kamel et al. [8].

corresponds to a lower neritic slope environment. Sidi Salem Formation is unconformably overlained by Abu Madi Formation.

1.2.3. Abu Madi formation (Late miocene)

Abu Madi Formation is composed of interbedded sandstones and mudstones. This formation was deposited as an incised valley fill at the end of the Late Miocene. The depositional environment of Abu Madi Formation changes from fluvial to lagoonal with marine influence increasing upward.

The Plio- Pleistocene sequences are represented by the following formations.

1.2.4. Kafr El Sheikh formation (Early–middle pliocene)

Kafr El Sheikh Formation consists of mudstone sequence with thin limestone and sandstone interbeds. This formation extends all over the Delta area. In the Nile Valley it is accumulated at outer neritic to upper bathyal water depths, following a major lower Pliocene transgression resulted in the flooding of the former Eonile canyon and transforming it into a long, narrow gulf far south Aswan [12]. Rizzini et al. [4] suggested that-the Kafr El Sheikh Formation is accumulated as neritic mudstones on the present onshore delta- and in a basinal setting offshore. They suggested that the sands incorporated into this formation are essentially storm sands. This formation is conformably overlain by El Wastani Formation.

1.2.5. El Wastani formation (Late pliocene)

El Wastani Formation consists of thick sand beds interbedded with thin clay beds which become thinner toward the top of the formation. This formation is transitional between the shelf facies of the underlying Kafr El Sheikh Formation and the coastal and continental sands of the overlying Mit Ghamr Formation. It shows large forests due to the progradation and could also contain deltaic deposits.

1.2.6. Mit Ghamr formation (Late pliocene)

Mit Ghamr Formation consists of thick layers of sand and pebbles which show clay interbeddings of limited thickness. The depositional environment of this formation is probably shallow marine to fluvial. It is a typical fill-up basin with coastal sands, coquina beds, clay and peat. It conformably underlies the Bilqas Formation.

1.2.7. Bilqas formation (Holocene)

Bilqas Formation is the top sedimentary cover of the Nile Delta province. It consists of sand interbedded with clay rich in pelecypod, gastropod and ostracod fragments. The clays contain many fragments of vegetal matter and peat deposits. The deposition occurred most likely in lagoons and brackish swamp, interrupted by beach sands.

During the last two decades, the Nile Delta and the offshore Mediterranean Sea have been intensively investigated by the petroleum industry. Till now over 36 TCF of gas have been discovered since drilling began in 1966 [13]. The geologic history of the Nile Delta created multiple source, reservoir and seal combinations. Main source rocks are believed to occur in Late Mesozoic and Oligocene to Miocene sedimentary succession.

2. Sampling and analytical techniques

A total of twenty gas samples were obtained through drill stem testing or modular dynamic testing from 20 wells located in the offshore area of the eastern and western Nile Delta fields namely, Abu Qir and North Abu Qir from the western Nile Delta while Temsah, Wakkar and Port Fouad representing the eastern Nile Delta-province and the different reservoir stratigraphic ages namely Miocene and Plio-Pleistocene.

The isotopic composition of twenty gas samples was determined in BGR (Bundesanstalt für Geowissnschaften und Rohstoffe) lab in Hannover. Gas chromatograph (GC Varian CP-3800) was equipped with a Flame Ionization Detector and a Thermal Conductivity Detector for separation of gas components (C1–C5) using three capillary columns (First, Silica PLOT 30 m×0.32 mm, Second, CP-Sil 5CB 30 m x 0.32 mm). Permanent gases (atmospheric gas components) were separated using three packed columns (Molsieve-12× 1.5 m ×1/8 in., Hayesep Q 0.5 m ×1/8 in. nickel, and a Hayesep T 0.5 m ×1/8 in. nickel). All six columns are placed in one oven, heated to 50 °C at the start of the analysis. After 10 min, temperature was increased at 10 °C/ min up to 180 °C, which is held for another 10 min. Finally the oven temperature was reduced to become 50 °C.

Measurements of δ ^{13}C and δ D on the hydrocarbon gases were conducted with a Thermo Finnigan Delta plus XL mass spectrometer. Gas components were separated on a gas chromatograph and injected into the mass spectrometer. Isotope values are reported in the δ notation in per mil (‰) relative to the common PDB and SMOW standards.

Data of analysis including gas chromatography, gas chromatography-mass spectrometry were conducted through Stratochem and Corex Laboratories (New Maadi, Cairo).

3. Results and discussions

3.1. Bulk geochemical characteristics of natural gas

The abundant composition of the natural gases is dominated by methane ranging between 92.72% and 99.96%. The natural gas wetness denoted by the formula C2+/∑(C1-C5 ×100 and cumulative contents of the higher hydrocarbons from ethane up to pentane range from 0.05% to 6.76%. According to Schoell [14] the natural-gas may be classified as wet if it exhibits a wetness index greater than 5%. The natural gas samples examined from the western Nile Delta can therefore be classified as wet gas, while the major number of gas samples from the eastern Nile Delta is classified as dry gas except the samples produced from the two wells PSE-1and DEN-1. Wet gas is thermogenic in origin but associated with condensate in most cases while the dry gas may have either a microbial or late mature thermal origin [14] or may be related to that microbial alteration occurring within the reservoirs through the preferential removal of the C3+ components [15].

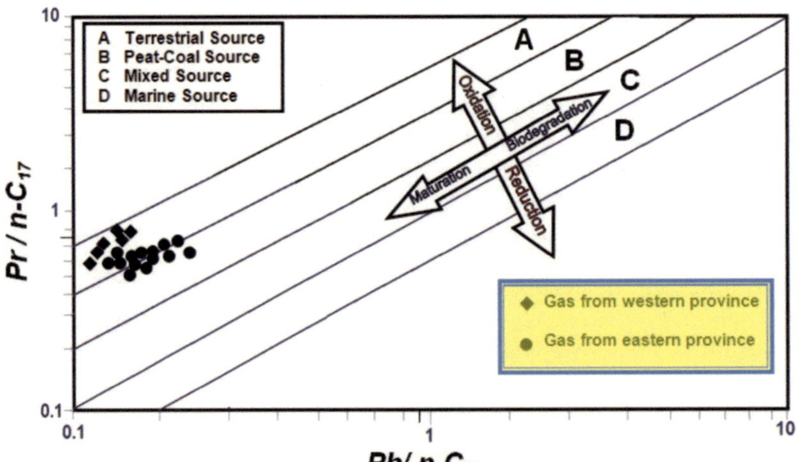

Figure 4. Relationship between isoprenoids Pr/n-C_{17} versus Ph/n-C_{18} showing source and depositional environments of the natural gas samples from the two provinces of the Nile Delta from Shanmugam[17]. All the study gas samples are located within the terrestrial to peat coal source depositional environments.

http://dx.doi.org/10.5772/60575

The predominance of pristane over phytane (Pr/Ph) ratio was used as an indicator of source rock depositional environments [16]. The ratio of Pr/Ph was ranged between 2 and 3 for the natural gas produced from the eastern province indicating fluviomarine and coastal swamp environment. The same ratio was found to be in the range between 6.4 and 9.4 which indicates peat swamp depositional environment. Figure 4 summarizes the achieved results concerning the nature of the source rock depositional environments is strongly supported by the plot of the isoprenoid ratios Pr/n-C_{17} versus Ph/n-C_{18} [17]. It can be seen from this figure that both the natural gases produced from the eastern and western provinces of the offshore Nile Delta and Mediterranean Sea are located within the region of terrestrial to peat coal source depositional environments. The high odd even carbon preference index (CPI > 1) for the studied natural gas samples from the two provinces reflect the generation from source facies dominating terrigenous and wax-rich components [18].

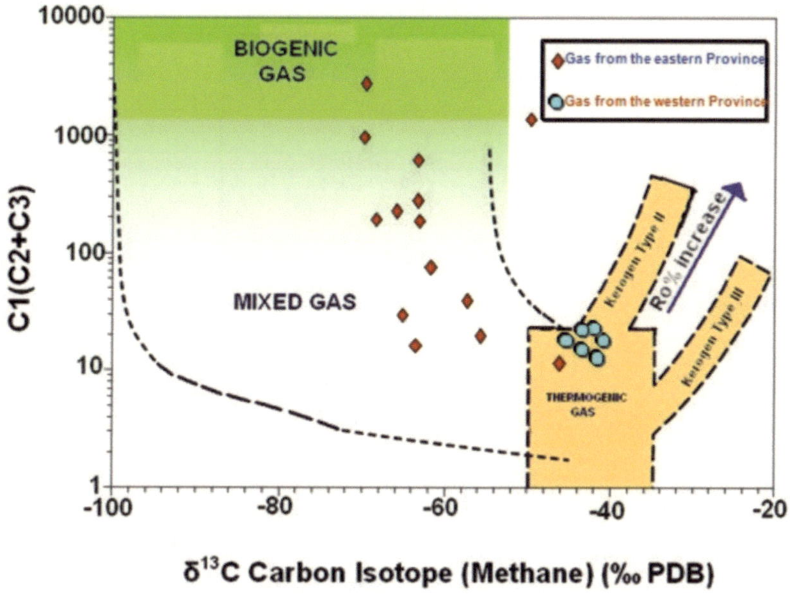

Figure 5. Genetic characterization of natural gases using carbon isotopes and C1 / C2 + C3 ratios modified after Bernard et al.[29] and Faber and Stahl [30].

3.2. Source and maturation dependent biomarker distributions

The results of organic geochemical analyses of twenty gas samples representing the western Nile Delta province namely, Abu Qir and North Abu Qir fields and the eastern Nile Delta province, namely Temsah, Wakkar and Port Fouad fields. The results of gas chromatogram mass fragmentogram of the study gas samples show a predominance of C_{29} stereoisomers > 50% over C_{27} regular steranes between 20% and 27% suggestings a significant input of terrestrial organic matter [19, 20]. Also, the achieved result is strongly supported by the relative abundance of diasterane index that ranges between 55% and 69% of the study gas samples.

All the study gas samples that exhibit higher oleanane index ranging from 19% and 42% strongly support the enrichment of angiosperm higher land plants input to the siliciclastic source kerogen which is thought to be derived from Late Cretaceous to Tertiary age [21, 22]. The medium concentration of moretane index in all the study samples ranges between 11% and 16% and the absence of gammacerane index strongly supports a terrestrial input [23, 24].

Biomarker maturity parameters, include the sterane isomerization ratios-C29 $\alpha\alpha\alpha$20S/(S+R), and C29 αßß/(αßß+$\alpha\alpha\alpha$) which according to Seifert and Moldowan [25] are genetically related to the effect of thermal maturity processes. The average sterane isomerization ratios $C_{29}\alpha\alpha\alpha$ $20S/(S+R)$ and $C_{29}\alpha$ßß/(αßß+$\alpha\alpha\alpha$) of natural gas are 0.6 and 0.5 respectively indicating a medium stage of thermal maturation equivalent to the main peak of oil generation window (0.85 Ro%) [18]. The medium sterane isomerization ratios may reflect the rapid rates of subsidence and sedimentation in the Nile Delta and appear to have been generated during the early stage of source rock maturation from type III kerogen [26].

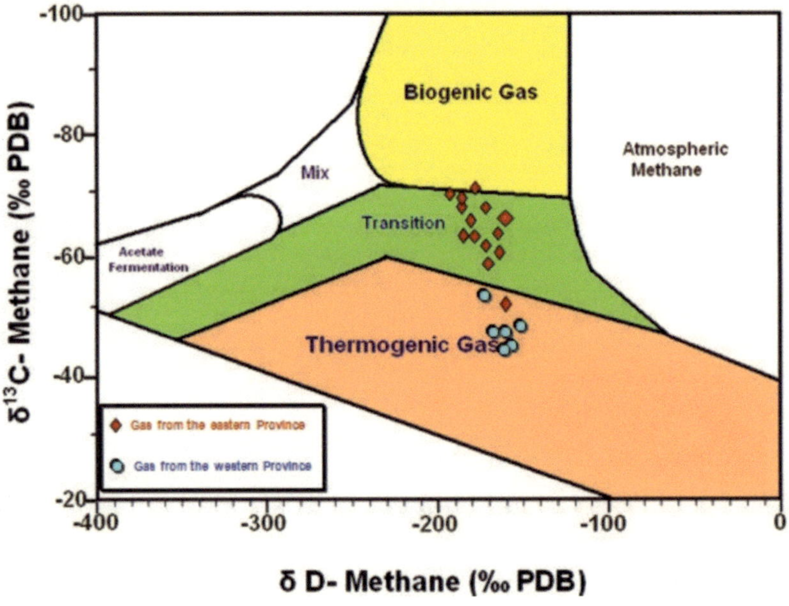

Figure 6. Carbon and Hydrogen isotope ratios of natural gases. Genetic fields according to Schoell[27, 31] and Whiticar et al. [32].

3.3. Relation between isotopic composition and origin of natural gas

The isotopic composition of natural gas depends on the type and maturity of source rock [27, 28]. Stable isotope geochemistry helps to differentiate the original gas derived from a single source rock. The methane carbon isotope compositions range between -65.6 ‰ and - 41.1‰. Methane hydrogen isotope data ranges from -193 ‰ to -149‰. The isotopically lightest δ^{13}C sample was obtained from the SEM-1 well, while the heaviest δ^{13}C value was received from

AQ-16 well. Many of the δ ^{13}C values have intermediate positions between these end members, which are consistent with what is commonly interpreted as mixed microbial and thermogenic gas [14, 28]. A general increase in the δ ^{13}C values with depth from Miocene to Plio-Pleistocene reservoir is consistent with the increasing thermal maturity and the decreasing relative concentration of biogenic gas.

The wet gas (C2+) components also display a significant variability in their isotopic compositions. The δ ^{13}C of ethane (δ $^{13}C2$) ranges from -37.5‰ to -27.9‰ but that of propane (δ $^{13}C3$) ranges from -32.2‰ to -11.1‰.

Natural gases become isotopically heavier and contain relatively more methane with increasing thermal maturity. Thermogenic methane is generally enriched in δ ^{13}C compared with microbial methane [27].

Figure 5 represents the Bernard plot in which the molecular ratio C1/ (C2 + C3) is plotted versus methane's stable carbon isotope ratio according to Bernard et al.[29] and Faber and Stahl [30]. The relationship between these two parameters allows distinguishing microbially generated methane from thermogenic hydrocarbon gases. The figure provides a strong indication for gas mixing of thermogenic and early microbial methane.

The δ ^{13}C methane versus δ D methane plot (Figure 6) represents a scheme for the carbon and hydrogen isotope ratios of methane and their genetic implications. Although this diagnostic plot was initially established for determining the thermogenic origin of methane [31] it is useful for the discrimination of microbial methanogenesis, i.e. carbon dioxide (CO_2) reduction versus acetate fermentation [32]. The δ ^{13}C values between -65.6 ‰ to -41.1‰ and δ D values between -193 ‰ and -149‰ obtained for the methane of the samples from this study classify methane as mixtures between thermogenic and biogenic origin. Figure 7 gives a good matching with the published regional gas data. The relatively heavy hydrogen isotopic compositions (Figure 7) suggest that the microbial gas proportions are probably generated through CO_2 reduction [31, 32].

In the offshore Nile Delta and Mediterranean Sea, the Oligocene to Early Miocene sedimentary sequence is considered as the primary source of natural gas and condensate characterized as mixed type II/III kerogen [8, 26].

The isotopic composition of the individual gas components is a function of thermal maturity of the generated source rock kerogen. Stahl [33] proposed an empirical relationship between methane carbon isotopic composition and vitrinite reflectance. Berner et al. [34] and Berner and Faber [35] illustrated a model between the carbon isotopic compositions of methane, ethane, and propane and source rock maturities represented by vitrinite reflectance measurements. In this study we used the isotope maturity models developed by Berner and Faber [35] to estimate the type and maturity of the precursor (source) material (Figures 7and 8). The geochemical characteristics of natural gas from the Nile Delta indicating a derivation from source rocks rich in terrestrial precursors [8, 36, 37]. Consequently, the admixture of light microbial methane to the thermogenic hydrocarbon gas takes place during secondary processes, such as migration, microbial oxidation, or mixing. These processes usually affect the gas properties within the reservoir [28, 40].

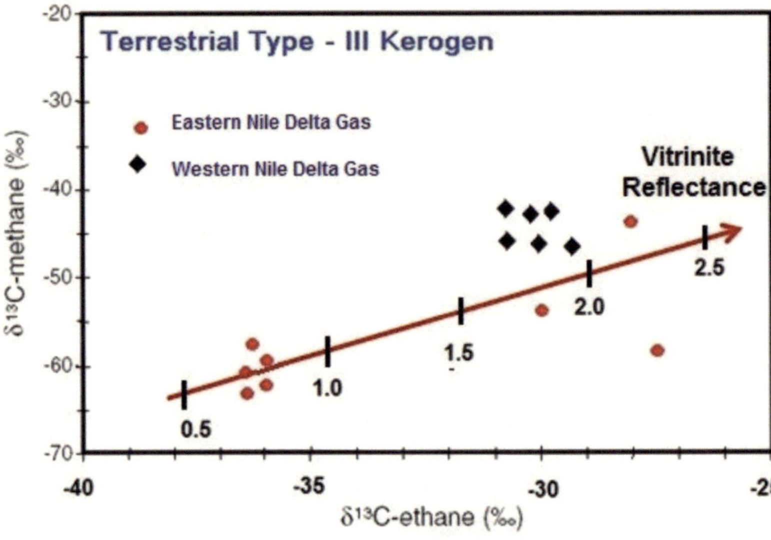

Figure 7. Isotopic composition of methane and ethane in natural gases of the western and eastern Nile Delta. The maturity line is according to Berner and Faber [35]. The terrestrial organic matters indicate the maturity of the source rock in vitrinite reflectance measurements

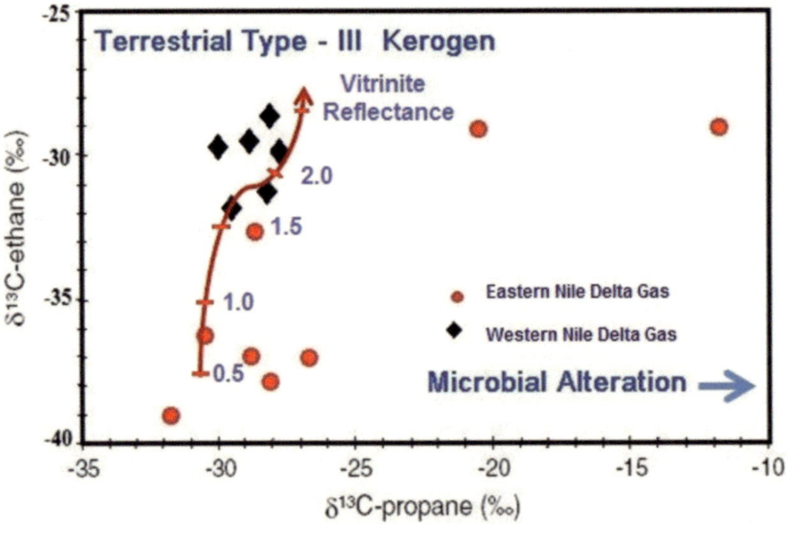

Figure 8. Isotopic composition of ethane and propane in natural gases of the western and eastern Nile Delta. The maturity line is according to Berner and Faber [35].

Isotopic compositions of ethane and propane from the west and east Nile Delta samples proved their mixing origin and derivation from terrestrial organic matter with a source rock maturity between 0.1% and 2.0% Vitrinite Reflectance measurements.

http://dx.doi.org/10.5772/60575

4. Summary and conclusions

The molecular and isotopic composition of twenty natural test gas samples produced from the Miocene and Plio-Pleistocene reservoirs of Abu Qir, N. Abu Qir, Temsah, Wakkar and Port Fouad fields representing the western and the eastern Nile Delta province were used to examine the generation and alteration of natural gas and condensate accumulations. The geochemical analyses revealed that the main constituents of the natural gases are methane 96.2–99.37% with minor contribution from ethane and propane. Methane carbon isotopic composition ranges from -65.6‰ to -40.3‰ PDB displaying a strong indication for gas mixing of thermogenic and early microbial methane. The gas chromatography_mass spectrometry carried out to the produced natural gases exhibit high oleanane index that ranges between (19% and 42%), a medium concentration of moretane index between 11% and 16% and the absence of gammacerane index and predominance of C_{29} stereoisomers > 50% over C_{27} regular steranes (between 20% and 27%) and a high diasterane index that ranges between 55% and 69%. The above mentioned results pointed out that the natural gases were derived from siliciclastic source rocks containing type-III kerogen of terrestrial origin and higher land plants input from Late Cretaceous to Tertiary age. The calculated maturity parameters based on various sterane isomerisation ratios, i.e.$C_{29}\alpha\beta\beta/(\alpha\beta\beta+\alpha\alpha\alpha)$and $C_{29}\alpha\alpha\alpha 20S/(S+R)$ reached to 0.6 and 0.5 respectively indicating a medium stage of thermal maturation equivalent to the main peak of oil generation window (0.85 Ro%). The medium sterane isomerization ratios may reflect the rapid rates of subsidence and sedimentation in the Nile Delta and appears to have been generated during the early stage of source rock maturation.

Acknowledgements

The author wish to express the deepest gratitude to the management of the Western Desert Operating Petroleum Company (Wepco) and Belayim Petroleum Company (Petrobel) for providing test gas and condensate samples to accomplish this work. Sincere thanks and gratitudes are also due to the Egyptian General Petroleum Corporation (EGPC) for granting the permission to publish this paper.

Author details

Mohamed Abdel-Aziz Younes*

Address all correspondence to: mohamedyounes@yahoo.com

Geology Department (Moharrem Beck), Faculty of Science, Alexandria University, Alexandria, Egypt

References

[1] Said R., (1962) The Geology of Egypt: Elsevier, Amsterdam, pp. 337

[2] Said R., (1981) The geology and evolution of the River Nile: Springer, New York, N.Y., 151 p.

[3] Ross, D. A., and Uchupi E., (1977) Structure and sedimentary history of southern Mediterranean Sea-Nile cone area: Am. Ass. Petrol. Geol. Bull., Vol. 61, p. 872–902.

[4] Rizzini, A., Vezzani, F., Cococcetta V., and Milad, G., (1978) Stratigraphy and Sedimentation of A Neogene-Quaternary section in the Nile Delta area, (A.R.E.). Mar. Geol., vol. 27, p.327–348.

[5] Zaghloul Z. M., EL Shahat A., and Hegah, O., (1980) Mineralogy of the Tertiary–Quaternary subsurface sediments, West Nile Delta area. Egypt. Jour. of Geology, vol. 24, p.177–188.

[6] Helmy, M., and Fouad, O., (1994) Prospectivity and play assessment of Abu Qir area, Nile Delta, Egypt. EGPC, 12th Petroleum Exploration and Production Conference, Cairo, vol. 1, 17, pp. 276–292.

[7] Sarhan, M., Talaat, M., Barsoum, K., Bertello, F., and Nobili, M., 1996. The Pliocene play in the Mediterranean offshore: Structural setting and growth faults controlled hydrocarbon accumulations in the Nile Delta basin. A comparison with the Niger Delta basin. EGPC, 13th Petroleum Exploration and Production Conference, vol. 1,7, p. 121–139.

[8] Kamel, H., Eita, T. and Sarhan, M., (1998) Nile Delta hydrocarbon potentiality. EGPC, 14th Petroleum Exploration and Production Conference, vol. 2, pp. 485–503.

[9] El Barkoky, A., and Helal, M., (2002) Some Neogene stratigraphic aspects of the Nile Delta. Mediterranean offshore conference. MOC, 2002.

[10] Dewidar Kh. and Frihy, O. (2007) Pre- and post-beach response to engineering hard structures using Landsat time-series at the northwestern part of the Nile Delta, Egypt : Journal of Coast Conservation vol. 11: pp. 133–142.

[11] Dewidar Kh. and Frihy, O. (2010) Automated techniques for quantification of beach change rates using Landsat series along the North-eastern Nile Delta, Egypt: Oceanograpic and Mar. Sci., Vol. 1(2) pp. 028-039.

[12] Chumakov, I. S., (1967) Pliocene and Pleistocene deposits of the Nile Vally in Nubia and Upper Egypt. Doki. Akad. Nauk. SSSR., p. 5–170.

[13] Dolson, J.C., Boucher, P.J., Dodd, T., Ismail, J., (2002) Petroleum potential of an emerging giant gas province, Nile delta and Mediterranean Sea o. Egypt. Oil and Gas Journal, 32–37.

[14] Schoell, M., (1980) The hydrogen and carbon isotopic composition of methane from natural gases of various origins. Geochimica et Cosmochimica Acta v.44, pp. 649–662.

[15] James, A.T., Burns, B.J., (1984) Microbial alteration of subsurface natural gas accumulations. Am. Ass. Petr. Geol. Bulletin v. 68, pp.957–960.

[16] Lijambach, G.W., (1975) On the origin of petroleum. Proceeding of the 9th World Petroleum Congress, Applied Science Publisher, London. 2, 357-369.

[17] Shanmugam, G. (1985) Significance of coniferous rain forests and related oil, Gippsland Basin, Australia. Am. Ass. Pet. Geol. Bulletin, 69, 1241-1254.

[18] Peters, K. E. Snedden, J. W., Sulaeman, A., Sarg, J. F. and Enrico, R. J. (2000) A new geochemical sequence stratigraphic model for the Mahakam delta and Makassar slope, Kalimantan, Indonesia: Am. Ass. Pet. Geol. Bulletin, 84, 12-44.

[19] Huang, W.Y. and Meinschein, W.G., (1979) Sterols and ecological indicators: Geochim. et Cosmochim. Acta, v. 43, p.739-745.

[20] Huang, H. P. (2000) The nature and origin of petroleum in the Chaiwopu Sub-Basin (Jungar Basin) N. W. China. Journal of Petroleum Geology, v. 23, pp. 193-220.

[21] Ekweozor, C. M., Okogun, J. I., Ekong, D.E.U. and Maxwell, J. R. (1979) Preliminary organic geochemical studies of samples from the Niger Delta (Nigeria). Analyses of crude oils for triterpanes. Chemical Geology, 27, 11-29

[22] Moldowan, J.M., Dahl, J., Huizinga, B. and Fago, F., (1994) The molecular fossil record of oleanane and its relation to angiosperms. Science 265, 768-771.

[23] Connan, J., Bouroullec, J., Dessort, T. D. and Albrecht, P. (1986) The microbial input in carbonate anhydrite facies of sabkha environment from Guatemala; A molecular approach In : Leythaeuser, D. and Rullkotter, J. (Eds.) Advances in Organic Geochemistry, 1985. Oxford-Pergamon, 20-50.

[24] Mann, A. L., Goodwin, N. S. and Lowes, S. (1987) Geochemical characteristics of lacustrine source rocks: A combined palynological - molecular study of Tertiary sequence from offshore China. In Proced. Indonesian Petroleum Association, 16th Annual Convention, Jakarta, 241-258.

[25] Seifert, W.K. and Moldowan, J. M. (1981) Paleoreconstruction by biological markers. Geoch. et Cosmoch. Acta, 45, 783-795.

[26] Sharaf, L.M. (2003) Source rock evaluation and geochemistry of condensates and natural gases, offshore Nile Delta, Egypt. Journal of Petroleum Geology, V.26(2), pp. 189-209.

[27] Schoell, M., (1984) Stable isotopes in petroleum research. In: Brooks, J.B., Welte, D. (Eds.), Advances in Petroleum Geochemistry, vol. 1. Academic Press, London, pp. 215–245.

[28] Whiticar, M.J., (1994) Correlation of natural gases with their sources. In: Magoon, L.B., Dow, W.G. (Eds.), The Petroleum System From Source to Trap, vol. 60. Am. Ass. Pet. Geol. Memoir, pp. 261–283.

[29] Bernard, B., Brooks, J.M. and Sackett, W.M., (1978) Light hydrocarbons in Recent Texas continental shelf and slope sediments. Journal of Geophysical Research vol. 83, 4053–4061.

[30] Faber, E. and Stahl, W., (1984) Geochemical surface exploration for hydrocarbons in the North Sea. Am. Ass. Pet. Geol. Bulletin vol.68, pp.363–386.

[31] Schoell, M., (1983) Genetic characterization of natural gases. Am. Ass. Pet. Geol. Bulletin vol.67, 2225–2238.

[32] Whiticar, M.J., Faber, E., Schoell, M., (1986) Biogenic methane formation in marine and fresh water environments: CO2- reduction vs. acetate fermentation – Isotope evidence. Geochim. et Cosmochim. Acta 50, 693–709.

[33] Stahl, W.J., (1977) Carbon and nitrogen isotopes in hydrocarbon research and exploration. Chemical Geology 20, 121–149.

[34] Berner, U., Faber, E., Scheeder, G., Panten, D., (1995) Primary cracking of algal and land plant kerogens: kinetic models of isotope variations in methane, ethane and propane. Chemical Geology 126, 233–245.

[35] Berner, U. and Faber, E., (1996) Empirical carbon isotope/maturity relationships for gases from algal kerogens and terrigenous organic matter, based on dry, open-system pyrolysis. Oganic Geochemistry 24, 947–955.

[36] Wever, H.E., (2000) Petroleum and source rock characterization based on C7 star plot results: examples from Egypt. American Association of Petroleum Geologists Bulletin 84, 1041–1054.

[37] Vandre′, C., Cramer, B., Gerling, P., Winsemann, J., (2004) The geochemistry of natural gas and condensates from the western o.shore Nile delta, Egypt. In: Proceedings of the DGMK Conference, Celle, Germany, pp. 371–380.

[38] Mosconi, A., Rebora, A., Venturino, G., Bocc, P., and Khalil, M. H., 1996. Egypt – Nile Delta and North Sinai Cenozoic tectonic evolutionary model. EGPC, 13th Petrol. Exp. Prod. Conf., Cairo, vol. 2, p. 203–223.

[39] Hemdan, K., El Alfy, M., Enani, N., Barrasi, M., and Monir, M., (2002) Structural Complexity of Pliocene and its Impact on Trapping Mechanism, N. Port Said Concession, Egypt. Mediterranean offshore conference. (MOC, 2002).

[40] Claudius, V., Bernhard, C., Gerling, P.and Winsemann, J. (2007). Natural gas formation in the western Nile delta (Eastern Mediterranean): Thermogenic versus microbial. Organic Geochemistry v.38, pp. 523–539.

Section 2

Synthesis and Processing

Chapter 3

Synthesis of Nanostructured Materials for Storing Hydrogen as an Alternative Source to Fossil Fuel Derivatives

Omar Reyes-Martínez and
José Luis Velázquez Ortega

Additional information is available at the end of the chapter

http://dx.doi.org/10.5772/61076

Abstract

The decline in global reserves of fossil fuels due to the increasing energy demand has caused the petrochemical industry to be interested in finding new energy sources. An alternative that should be considered is the technology based on hydrogen as it is friendly to the environment and its combustion is greenhouse.

The technology based on hydrogen considers the synthesis of metal-organic materials that have the ability to store, although this phenomenon (storage) is one of the biggest problems to overcome. For this reason, options synthesized for storing materials are provided in this chapter of Book Two. These materials are considered nanostructured systems with specific features in 2D and 3D. The synthesized materials are tetracyano-niquelates ([M $(H_2O)_2$ (Ni $(CN)_4$) $(H_2O)n$ M = manganese, cobalt, and nickel) and hexacyano-cobaltates ((M_3 [Co (CN) $_6$] $2xH_2O$); M: manganese, nickel, cobalt, zinc, cadmium, and copper), both systems with individual cavities that require study and characterization. Moreover, in this chapter, a thermogravimetric characterization is performed in order to establish the activation energy of the mobility of water molecules contained in the cavities, relevant information to establish the energy barrier to overcome in the storage of hydrogen.

Keywords: Fossil fuel, hydrogen, nanoporous, thermogravimetry, activation energy

INTECH
open science | open minds

© 2015 The Author(s). Licensee InTech. This chapter is distributed under the terms of the Creative Commons Attribution License (http://creativecommons.org/licenses/by/3.0), which permits unrestricted use, distribution, and reproduction in any medium, provided the original work is properly cited.

1. Introduction

In recent years, micro-porous and nanocomposites have been wide, and there are special interests in specific areas in order to generate improved technological processes such as separation, gas adsorption, ion exchange, and heterogeneous catalysis, to mention some of them [1, 2].

Particularly in the separation and gas adsorption area in microporous and nanosystems, research has focused on hydrogen's adsorption and storage as an alternative energetic alternative resource since this element might become a long-term fossil fuel substitute because its caloric power is three times higher than gasoline (142 kJ/mol). It is noteworthy that when the hydrogen is at room temperature in its supercritical state ($Tc = 32.7\,°K$), storage is difficult, hence the importance of finding new porous materials having cavities with the ability to store it safely. Similarly, once stored, it can be used in mobile systems (automotive), avoiding the production of greenhouse gas emissions [3].

Studies on porous media properties have focused mainly on inorganic materials such as zeolites or carbon-based materials such as activated carbons, the last one having a high porosity on one side and an irregular pore, while zeolites have opposite characteristics [4].

In order to generate a material having defined pores and specific physicochemical properties, the interest is held in the synthesis of new porous materials showing specific conditions from defined synthesis models, examples of these are as follows: (i) layered systems (2D), tetracyano-niquelates attached to a metal transition [$M\ (H_2O)_2\ (Ni\ (CN)_4)\ n\ (H_2O)$], M = Co, Ni, Mn, known as Hoffmann-type compounds; and (ii) latticed porous materials, hexacyano-cobaltates bound to a transition metal [$M_3\ [Co\ (CN)_6]\ 2\ nH_2O$], M = Co, Ni, Mn, Cd, Cu, Zn, known as Prussian blue analogue materials.

The importance of lamellar materials known as two-dimensional (2D) lies in the ability to include molecules in their interlamellar spaces known as guest molecules, with the sole purpose of generating pores or cavities with specific characteristics according to the same properties of the included molecule, generating a three-dimensional network, in addition, to be able to store various molecules such as H_2 and CO_2 in the pores obtained. The transformation of a layered structure in a porous structure throughout a molecular spacer insertion was first introduced into the clay mineral smectite to overcome the limitations of the size of the cavities in zeolites for the new materials obtained [5].

The proposed 2D systems have water molecules between the sheets, joined by the system for various attractive forces. There are two types of water molecules included in the material, some are called coordinated water molecules, which are forming bonds with transition metals present in the sheets and play the role of guest molecules to form the pillars of the final structure. Additionally, there are others known as zeolitic water, which are placed in the cavities or interlamellar region linked by hydrogen bond type interactions.

The water molecules (coordinated and zeolite) are also in the hexacyano-cobaltates, but the interaction with the system is weaker so the energy required to abandon the system is lower compared with the 2D systems.

Another important feature of layered systems is its ability to crystallize into three different phases known as L0, L1, and K [6, 7, 8]. All the structures contain water molecules between blades, and each phase has a different number of water molecules. The structural configuration that adopts the set of water molecules and its interaction with the crystal lattice is the cause of the three different phases.

Latticed porous materials are crystallized in a cubic cell (Fm-3m) and have between 10 and 13 water molecules (coordinated zeolite) per unit cell. In these structures, the transition metals (Co, Mn, Cd, Zn, Cu, and Ni) have the center of inversion, and Ni atoms are in a plane of symmetry [7]. Known latticed porous materials (hexacyano-cobaltates) as Prussian blue analogue materials are interesting to form windows and pore volumes suitable for the separation and storage of small molecular forms, as in the case of the hydrogen molecule [9, 10]. In both families, the main objective is to generate materials that provide these properties and specific characteristics such as pore size, shape of the window, pore volume of adsorption, etc., as this will dictate whether it is a functional material to be used as molecular sieve to catch species among its pores.

Once the structures of porous materials are defined, the property characterization of these is critical because it will determine whether the guest molecules may or may not remain in such pores. For this reason, special emphasis must be taken on the study of the interactions of the water molecules (coordinated and zeolitic water molecules) partially hosted on both materials and kinetic parameters involved in the dehydration process.

In the study of kinetics and their associated parameters, one of them is activation energy (Ea), which is the most relevant to study without leaving physicochemical aspects such as thermal stability and structural systems obtained from different aspects. There are several forms of useful characterization to define the structures and properties of materials under study. The use of thermal analysis is a fundamental technique, especially supported by high-resolution modulated thermogravimetric analysis (Hi/ResMTGA) supplemented with the technique of X-ray diffraction (XRD), which is the immediate basic characterization to provide relevant information and thus provide the monitoring structural changes during dehydration processes.

2. Porous systems

At the end of the 1990s, there were basically two types of porous materials: inorganic and carbon derivatives. In the case of the former, the two most important subclasses of materials with open structures are aluminosilicates (zeolites) and aluminophosphates (AlPO4s); whose crystalline lattices offer permanent porosity to be stable before the removal of molecules of water of hydration. The market of zeolites is of several million tons per year, and it has success in a wide range of applications, especially in the petrochemical industry and the hydrocarbon separation, purification of gases and liquids, and the catalytic fragmentation of chain hydrocarbons long. Additionally, they have also been applied in ionexchange as detergent additives and the separation of gases and solvents, for example, as "molecular sieves" to the dehydration of organic solvents.

In the case of activated carbons, they have large specific surface area as well as high porosity; however, they have a disordered structure. These materials are widely used in the processes of separation, catalytic converters, capacitors, storage of gas, and biomedical engineering applications [11].

2.1. Zeolites

Zeolites have been widely studied because of their H_2 storage capacity. For equal surface areas, there is a smaller capacity of absorption in the case of zeolites in comparison with the ones which are bases on carbon structures.

Zeolites have structures mostly mesoporous, with volumes of relatively small microporous, which make them the least promising materials based on H_2 storage, as referred to in [12]. Their abilities range from 2% to 2.5% based on the weight of the lattice. Zeolites are crystalline aluminosilicates with a structure consisting of a three-dimensional combination of TO_4-tetrahedra (T =Si, Al) linked together via oxygen atoms. The structural formula of zeolites can be expressed as [13] $M_{x/n}$ *[(AlO*$_2$*)*$_x$ *(SiO) and]. wH*$_2$*O*, where *M* is the exchangeable cation, *n* is the cation's valence, (*x*, and) are the total number of tetrahedra per unit cell, and *w* is the number of water molecules. The structure of zeolites presents channels and cavities of molecular dimensions, which are water molecules, adsorbates, and compensating cations' charge (negative charge created by the presence of AlO_4 structure). These channels and cavities give zeolites a porous structure, which allows these materials to have a very large internal surface compared to the external.

Some zeolitic materials pass through an intermediary laminating these precursors during their training, evidenced by their X-ray diffractogram. Zeolitic laminars are attractive candidates for pillarization processes, which could result in very interesting features such as microporous sheets and activity type zeolite, together with their properties such as mesoporous adsorbents.

Leonowicz et al. [14] proposed two sets of independent pores for the so-called zeolite MCM-22 type. One of the systems of pores is defined by sinusoidal and bidirectional channels; the other consists of large supercavities with a 7.1-Å free inner diameter *A* and an internal height of 18.2 Å. Figure 1 schematically illustrates the structure of MCM-22, where you can see the two systems of pores.

2.2. Structures of carbon-based materials

Materials with carbon-based structures offer potential to adsorb H_2, values of relatively low density, appropriate chemical stability, and large pore structure, and they can be found in a wide variety of structural forms. These forms are closely related to the conditions of synthesis, carbonization, and activation, employed during its preparation [15]. Among the materials based on carbon are the so-called activated carbons (AC). In this type of material, the existence of a porous structure is determined by the spatial arrangement of the grapheme, which can be

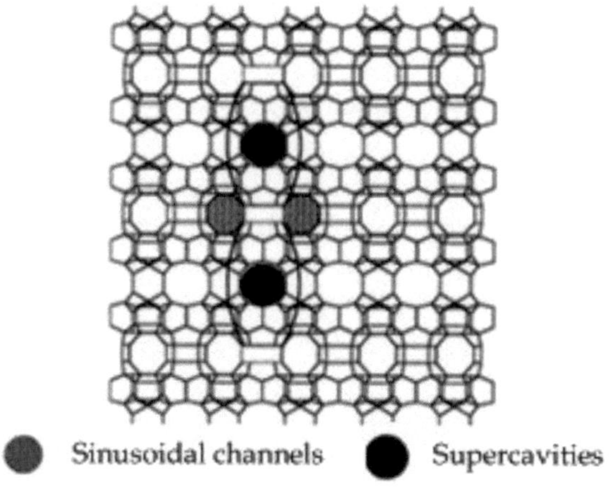

Figure 1. Diagram of the zeolite MCM-22 structure [1].

stacked to give place to the development of a porous structure relatively little polar. The adsorption/desorption processes of H_2 in such materials are characterized by a relatively rapid kinetics, and corresponding isotherms are hysteresis, in which results are attractive in systems that require a high-speed H_2 loading and unloading. However, the use of these materials is limited by highly dependent temperature and pressure adsorption capabilities [16, 17].

In materials based on structures of carbon, whose surface is not chemically modified, the physical adsorption of H_2 is only due to the existence of weak van der Waals adsorbent–adsorbate interactions. At environmental temperature, the energy corresponding to these interactions is similar to the energy of the thermal motion of the molecules of the gas, so by only decreasing the temperature, the adsorbent–adsorbate interaction energy becomes greater than the thermal movement, which decreases proportionally with the temperature [18, 19].

It has been reported that at low pressures, the amount of adsorbed H_2 increases with the increasing density in carbon nanostructures because the pores favor narrowing H_2–surface interactions. At high pressures, the specific surface area available for the adsorption of H_2 is the determining factor, and the amount adsorbed increases by decreasing the material density [20].

2.3. Layered systems

Lamellar materials are solids with two-dimensional structures possessing atoms firmly linked together in two directions of space forming foil (planes), and they are weakly linked in the perpendicular direction among them. The region of weak interaction between the sheets is normally known as interlamellar region or gallery [21]. Many kinds of solid lamellar blades have electric charge due to substitutions of isomorphic ions of different network load. To reach the electrical neutrality of the structure, ions of opposite charge, normally solvated by water or other polar molecules, occupying the interlamellar region, are known as compensation ions. Layered solids have high surface areas between 100 and 1000 m^2/g. In many cases, the interlamellar surface is accessible only to water and other small polar molecules that are capable of solvated ions of compensation.

The elimination of molecules of solvation by degassing at elevated temperatures results in a collapse of the interlamellar region, especially if interspersed compensating charge ions are small in relation to the occupied space; on the other hand, if the compensating charge ions are relatively large, they can have the role of pillars and prevent the collapse of the interlamellar regions when the middle of solvation is eliminated, resulting in solids called solid lamellar pillars. Figure 2 presents a system that shows some characteristics of layered solids.

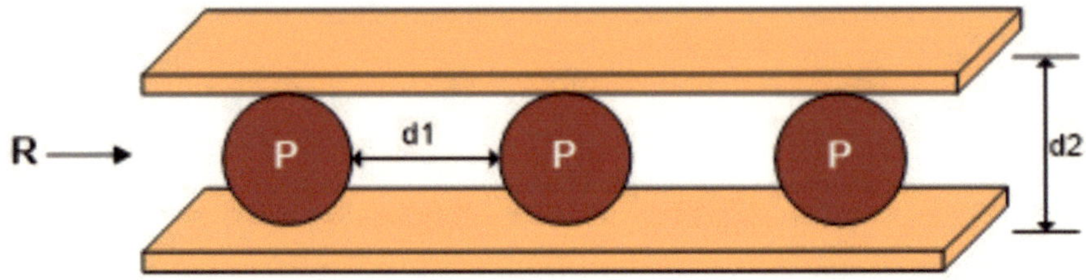

Figure 2. Representation of a layered solid pillared, wherein pores are defined by the height of the interlamellar space (d2) and the lateral spacing (d1) between the guest molecules or pillars (P). R denotes the interlamellar region.

Laminar solid pillars can be described as sandwich compounds serving three important criteria:

1. Interlamellar species must be sufficiently robust to promote the vertical expansion of the blades (d2) and to prevent their collapse during any thermal process to which they are subjected.

2. The pillars must be sufficiently separated (d1) to allow access to molecules. Simply expanding the blades to molecular size, collation of pillars, is not significant in relation to the properties of adsorption and catalytic behaviour of solid if the interlamellar region is fully occupied by pillars.

3. The blades must be rigid enough to maintain the separation desired between pillars. Otherwise, the flexibility of the blades could close gaps between pillar and pillar.

Lamellar surface area presented by solids (between 100 and 1000 m^2/g) makes these materials an object of study important above all in the area of nanoscience and nanotechnology because they have been shown to generate opportunities in different fields of application, such as biomedical electronics, applications in the area of storage and energy conversion, and catalysis [22].

2.3.1. Host molecules (pillarization)

The basic function of the pillarization is to include molecules in the interlamellar region to obtain a three-dimensional porous system with well-defined and specific characteristics according to the material or molecules that are embedded as a result. The interaction host species with active sites, which are normally present on the surface of the plates, is the driving force of the mechanism of formation of pillars in a structure laminating. These sites may be crystallographic positions or groups influenced by positive or negative charges, or sites with acid, basic, or polar character in general. The density of active sites on the surface on one side of the blade is expressed by the number of sites per cm^2 of surface layer.

The reciprocal value of the density of surface represents the average area per active site. This value, also called free area, is an important characteristic of a laminated solid. Knowing this, it is possible in many cases to predict whether the full collation of guest molecule stoichiometry is viable or not. In nature, the surface density and availability of the active sites is usually the same on both sides of the layers. However, in some cases, there are asymmetric blades.

The layout of the active sites in each side of the blade and its free area can be determined precisely when the blade structure is known.

In general, the pillarization process variables are as follows [23]:

1. Properties of the material, such as composition and location of the load on the blade
2. Nature of the agent formed with pillars
3. Ion exchange process
4. Subsequent treatments such as drying, cleaning, etc.

The most important parameters affecting the formation and the properties of the polynuclear cations as agent formed with pillars are as follows [24]:

a. Metal ion concentration
b. Basicity or degree of hydrolysis
c. Preparation temperature
d. Time and temperature aging
e. Method of preparation

The water molecules in some materials tend to be regarded as pillars of the same due to interactions that generate with the transition metals present, and the study of these molecules provides information of the nature of the material.

Pillar materials include clay Al-montmorillonite and the Al-bentonite, which is used in gas-oil cracking and reactions of phenols hydroxylation reactions, respectively.

2.3.2. Tetracyano-metallates

In recent years, great efforts have been devoted to research directed toward the synthesis and characterization of multidimensional metal complexes with cyanide bridge groups [25]. The most common and controlled strategy at the time of preparing this type of system is in autoassembly specifically designed predecessors. They are normally used as a complex cyanometallate that acts as a ligand and a complex of a metal transition coordination position for atoms of nitrogen from the groups free of cyanide (see Figure 3).

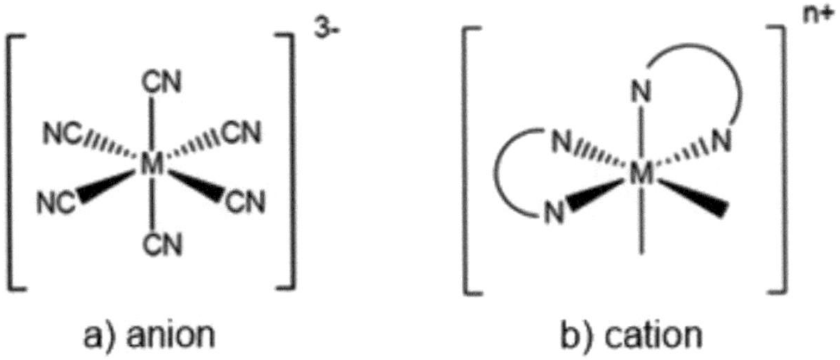

Figure 3. Assembly blocks of cyanometallate systems: (a) the anionic block [M $(CN)_6]^{3-}$, (b) the cationic block $[M'L_2]^{n+}$.

Tetracyano-metallates and hexacyano-metallates (Prussian blue analogues) are part of the materials with cyanide bridge group. Anion cyano (CN-) is a functional group with the ability to be simultaneously coordinated by both atoms at different cations. This makes it an excellent molecular block to generate different lattices structural at 1, 2, and 3 dimensions. The chemical bond in the cyanometallate in networks is the linear bridge M-CN-M; M = metal, with a distance M–M approximately 5 to 6 Å.

When linear polyhedron [Ag (CN)] acts as a binding between two centers of greater coordination, the others are connected to 2D or 3D networks [26]. In the union of blocks, the same metal center generates various structural patterns, showing the variability of crystal structures. There are other types of binders such as NH_3 and H_2O, which play an important role in the 3D structure stabilization and the volume of empty space that is occupied by species guests.

The initial structure of compounds including cyanometallate [Ni $(CN)\ _4]^2$. The flat square is the clathrate of Hoffmann's formula [Ni $(NH_3)\ _2$Ni $(CN)\ _4]$, where a 2D network shares a corner in units of [Ni $(CN)\ _4]^2$, separated by NH^3 binders. The resulting structure is in 2-dimensional (2D) structure, with the distance between blades occupied by the guest of benzene molecules that are perpendicular to the blades.

Materials with transition metals have been developed, which provide specific features to the two-dimensional network for the sole purpose of including pillars to further the creation of the same form, providing structures with defined pore and known dimensions. Such is the case of the tetracyano-niquelates. Although they are based on the principle of a structure-type Hoffmann, including transition metals such as Co, Mn, and Ni, the result is to have water molecules in the central part or interlamellar, which serve as pillars of the main structure.

The formation of tetracyano-niquelates involves square plane anion [Ni (CN) $4]^2$, with four sites attached to metal ions M = Ni, Co, Mn, via the nitrogen atoms of the cyano group. To have the octahedral geometry, the block has two further binders. For this particular case, those additional ligands bind to water molecules to complete the coordination sphere as shown in Figure 4. The consecutive joining molecular building blocks of this two-dimensional space can build a two-dimensional structure.

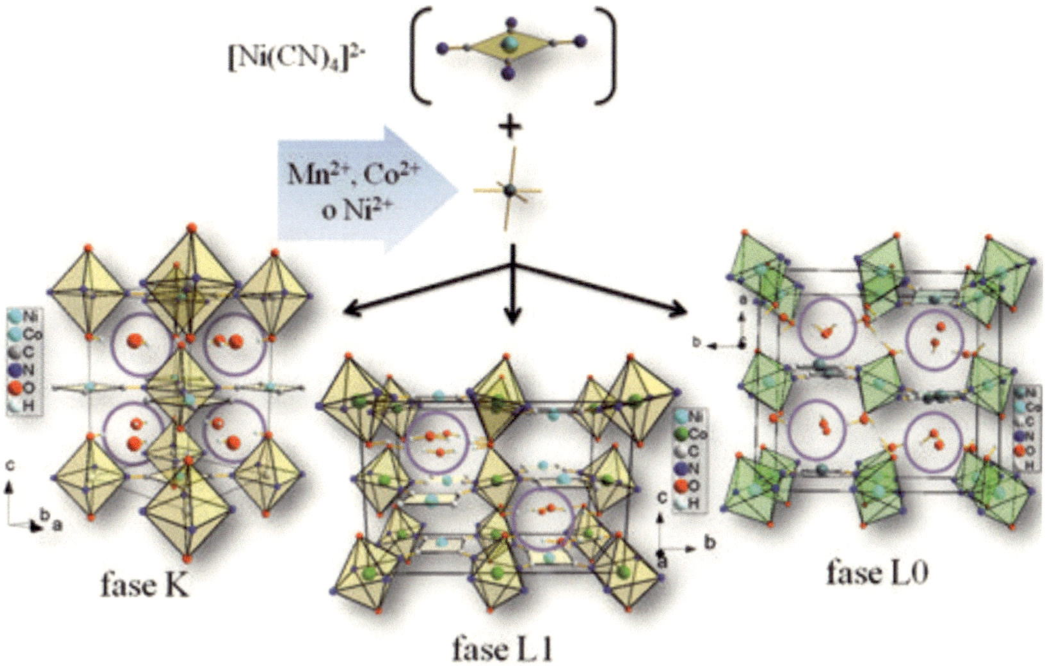

Figure 4. Schematic representation of the assembling process for obtaining K, L1, and L0 phases. The zeolite-like water molecules are highlighted with open circles in the interlayer region.

2.4. Prussian blue analogues

Historically, the "iron blue" is known as (ferric ferrocyanide) "soluble" Prussian blue and the blue is known as "insoluble" Turnbull (ferrous ferrocyanide); however, all recent studies conclude that both are identical. The formation of this pigment can be written as

$$4\left[\mathrm{Fe}\right]_{(ac)}^{3+} + 3\left[\mathrm{Fe}\left(\mathrm{CN}\right)_6\right]_{(ac)}^{4-} \rightarrow \text{Azul de Prusia, } \mathrm{Fe}_4\left[\mathrm{Fe}\left(\mathrm{CN}\right)_6\right]_3 x\mathrm{H_2O} \ \left(x = 14 - 16\right)$$

Its low solubility evidence is the polymeric nature of the solid product. Due to the 4:3 ratio of Fe (III) and Fe (II), 25% of the sites of Fe (II) are vacant. The occurrence of more than one oxidation state of the same element in a particular compound is defined as a mixture of valence. Due to the partial relocation of electrons from valence, the Prussian blue is a semiconductor.

Depending on the properties desired in the product, many routes of preparation are known. Given its porous nature, it has a certain zeolitic character, which allows you to store small molecules in their cavities [27].

They have now developed materials under the same principle of formation of Prussian blue by changing the formation of anionic block metals. The hexacyano-cobaltates (of the family of the hexacyano-metallates) are an example of these materials considered as Prussian blue analogues.

2.4.1. Hexacyano-metallates

The hexacyano-metallates of the transition metals are microporous molecular family, whose structures are based on a three-dimensional arrangement of chains M-C≡N-M. Metal M (inner metal), linked to the carbon atom, is always forming octahedral molecular blocks $[M^{n+} (CN)_6]^{6-n}$ as the metal M (external metal) acts as an assembler of the octahedral blocks and possibly octahedral or tetrahedral coordination [28, 29].

The first studies of H_2 adsorption in this polymer-coordinated family were published in 2005, simultaneous with the reports of Long et al. [30] and Kepert et al. [31], who studied the formula of general Prussian blue analogues M_3 $[Co (CN)_6]_2$, known as hexacyano-cobaltates as the internal metal cobalt and being able to change the external metal; M= Mn, Fe, Co, Ni, Cu, Zn, and Cd.

Figure 5 shows the structure of a Prussian blue analogue in the cubic phase due to the octahedral coordination adopted by the external metal.

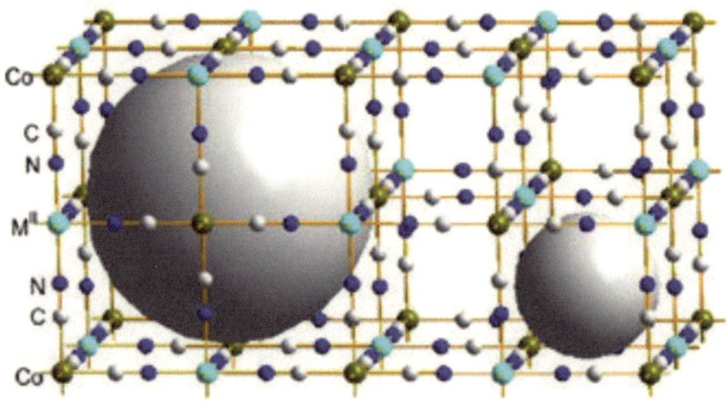

Figure 5. Porous framework for the cubic form, by a hexacyano-cobaltates and a transition metal [30].

2.5. Materials synthesis

Interest in synthesizing nanostructured materials has led to the study of families with specific characteristics. Here two families are addressed: (1) the family of tetracyano-niquelates (II) [with three different external metals, that is, (Co^{+2}) cobalt, nickel (Ni^{+2}), and manganese (Mn

$^{+2}$)], which is synthesized in L1, K, and L0 phases for three metals, and (2) the hexacyano-cobaltates (III) [with six different external metals, that is, cobalt (Co^{+2}), nickel (Ni^{+2}), manganese (Mn^{+2}), copper (Cu^{+2}), (Zn^{+2}) Zinc, and cadmium (Cd^{+2})]. X-ray diffraction (XRD) should be used to check the phases and structures of these materials.

Modulated thermogravimetry (MTG) is used for the kinetic analysis of dehydration, and scanning electronic microscopy (SEM) is used to analyze the morphology of the materials during the process of dehydration.

2.5.1. Tetracyano-niquelates synthesis (lamellar materials)

Figure 6 shows the procedure for the synthesis of materials [M (H_2O) $_2$ (Ni (CN) $_4$) (H_2O) *n*: M = Ni, Co, Mn, *n* = 1, 2, 4], which are obtained by the method of precipitation.

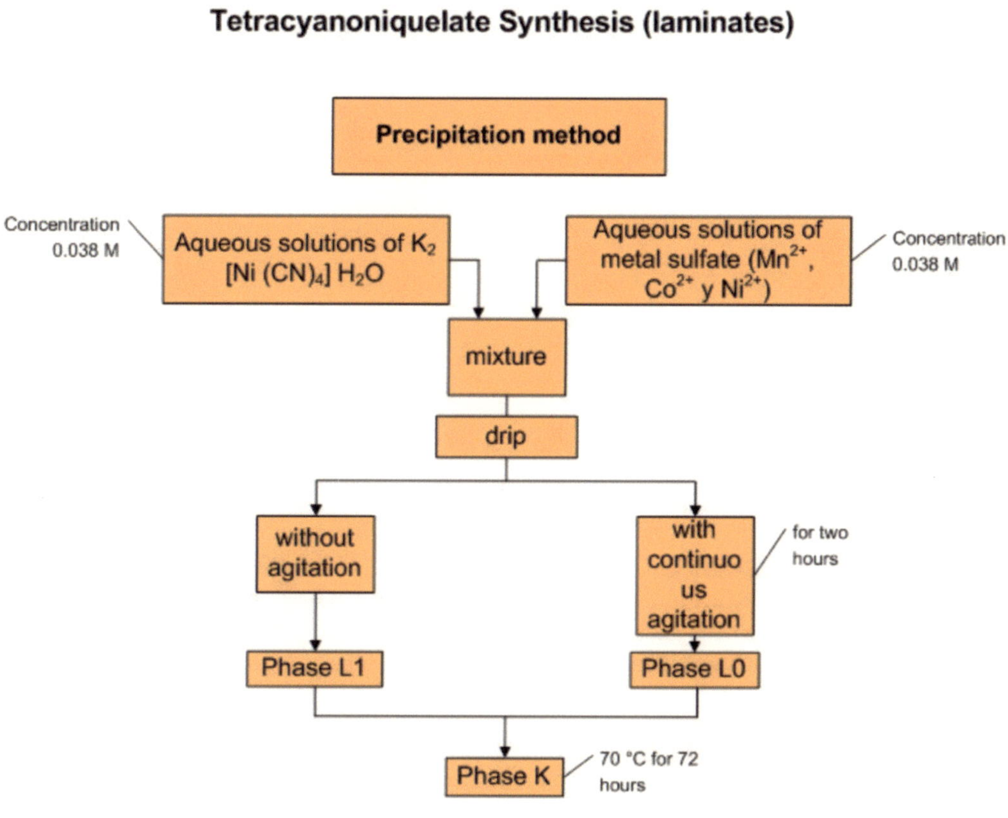

Figure 6. Materials synthesis $[M(H_2O)_2(Ni(CN)_4)n(H_2O)$: M = Ni, Co, Mn; n=1,2,4$]$.

Once the corresponding blends are obtained, the resulting precipitated solid is separated by centrifugation, and the solid fraction obtained is repeatedly washed with distilled water to remove the excess ions and then left at room temperature air-dried to constant weight.

2.5.2. Hexacyano-cobaltates synthesis (porous materials)

Figure 7 shows the hexacyano-cobaltates synthesis process.

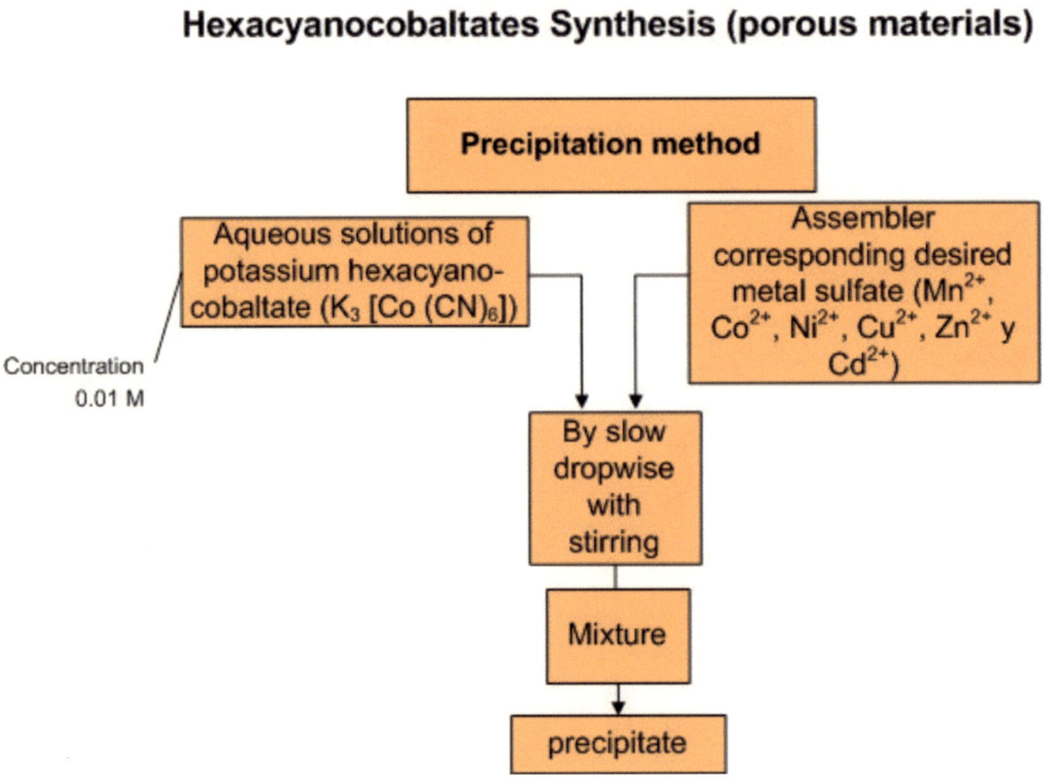

Figure 7. Hexacyano-cobaltates synthesis.

The amount in excess of the metals involved guarantees the formation of a single phase. The resulting precipitate remains for a week in the mixture. Then it is separated by centrifugation. The solid fraction is repeatedly washed with distilled water to remove the excess ions, and finally the material is air-dried to constant weight.

2.6. Characterization techniques

To ensure that the phases that have been obtained during the synthesis are correct, various techniques are used. This chapter will talk about three of them, which allude to the obtaining of the phases. One of them is X-ray diffraction (XRD), which is used under two aspects: to get the diffraction patterns of the synthesized material and to ensure that the crystalline phases of the structures meet the required specifications, which includes synthesis. On the other hand, this technique is used to perform a constant monitoring during the dehydration of materials and thus observe the structural change to drop out of the same water molecules present in the

http://dx.doi.org/10.5772/61076

system. Coupled with this technique, an SEM was used in order to follow up on the morphology of the material during the synthesis when dehydration was carried out. Finally, a thermal analysis to corroborate the number of molecules of water, in addition to providing information about the kinetic parameters, was involved during the dehydration of the study materials.

2.6.1. Systems lamellar characterizations

The family of tetracyano-niquelates (M $(H_2O)_2$ [Ni $(CN)_4$] $\cdot xH_2O$); M: Mn, Co and Ni, x = 1, 2, and 4 has been studied recently due to its ability to store molecules such as H_2, CO_2, and N_2. It is known that they crystallize in three different ways, that is, phase L1, phase K, and phase L0 [7]. The forms adopted due to the water molecules present in the structure and their interaction with metals assemblers. Figure 8 shows three phases reported.

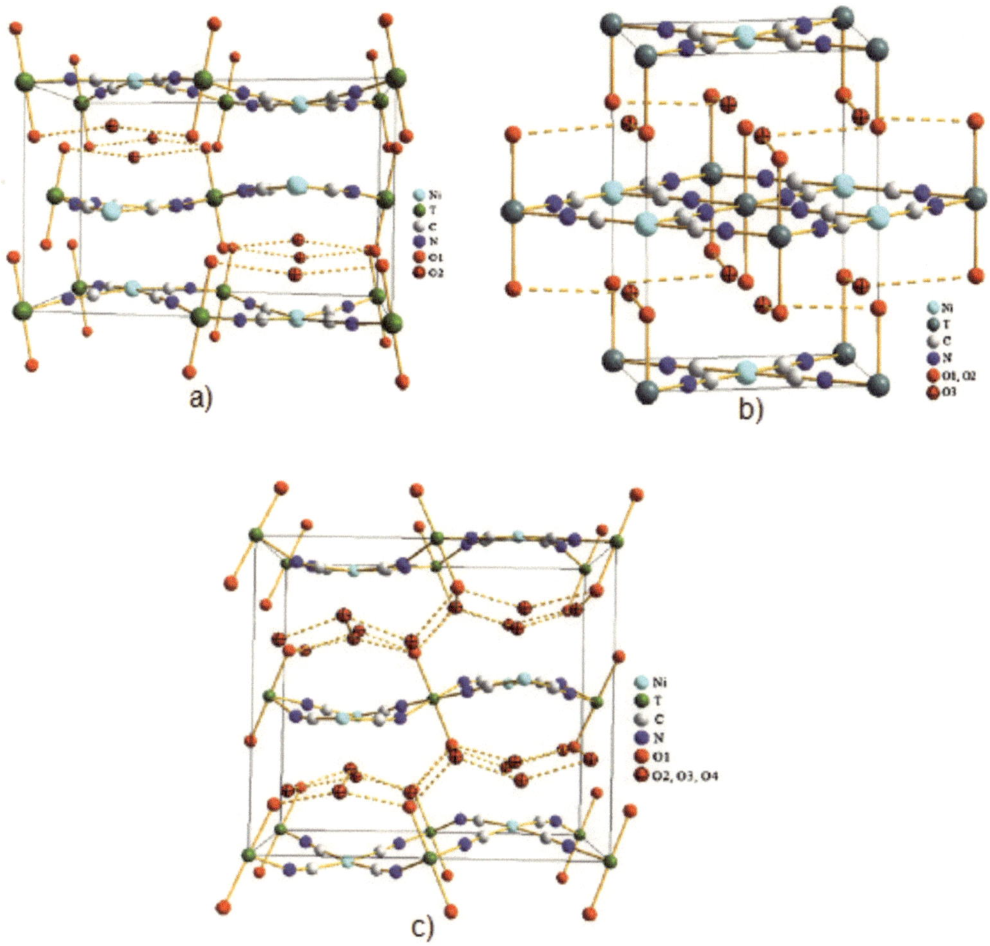

Figure 8. Framework for L1 (a), K (b), and L0 (c) phases. Coordinated water molecules from neighboring layers remains linked through aquo-bridges (hydrogen bonding interactions).

Thermal analysis is performed in four aspects: (i) analyze how to change kinetic parameters in a same phase for different metals of assembly, (ii) observe the influence of kinetic parameters over the same metal assembly on three different stages, (iii) analyze L0 to L1 phase change during the process of dehydration, and (iv) vary kinetic parameters as the partial pressure of water in the system is changed. All materials are needed to monitor the process of dehydration with the help of XRD to observe when there is a change in the structure or collapse once the material molecules are removed totally or partially.

2.6.1.1. Two-dimensional network of water molecules in the three phases interpretation

In order to understand how water molecules are found in the material and thus to make one website speedily of the kinetic parameters during the process of dehydration, two-dimensional networks of water molecules are presented in three different phases (see Figure 9). Note that the water molecules are organized along two-dimensional planes alternating the position between coordinated octahedral (defined by the position of the metal) sites and the position in the existing cavities between polyhedral.

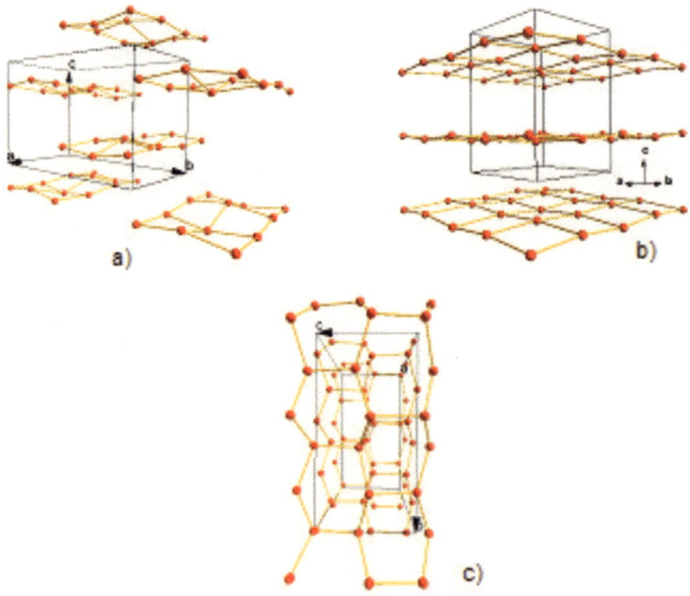

Figure 9. Two-dimensional representation of the network of water molecules in the different lamellar systems for (a) L1 phase, (b) K phase, and (c) L0 phase. Each red sphere is a water molecule without the hydrogen atoms.

2.6.1.2. Systems thermogravimetry lamellar characterization

Figure 10 shows samples of thermograms of the corresponding phase L1 as well as profiles of conventional or normal activation energy obtained from synthesized, likewise in gray color. The samples shown are the activation energy profiles obtained after dehydration of the corresponding material in its phase L0. The difference between the profiles of activation energy is due to local variations of the chemical potential of water or kinetic disabilities in reversible reactions caused by the removal of part of the water when the temperature is increased in the same way.

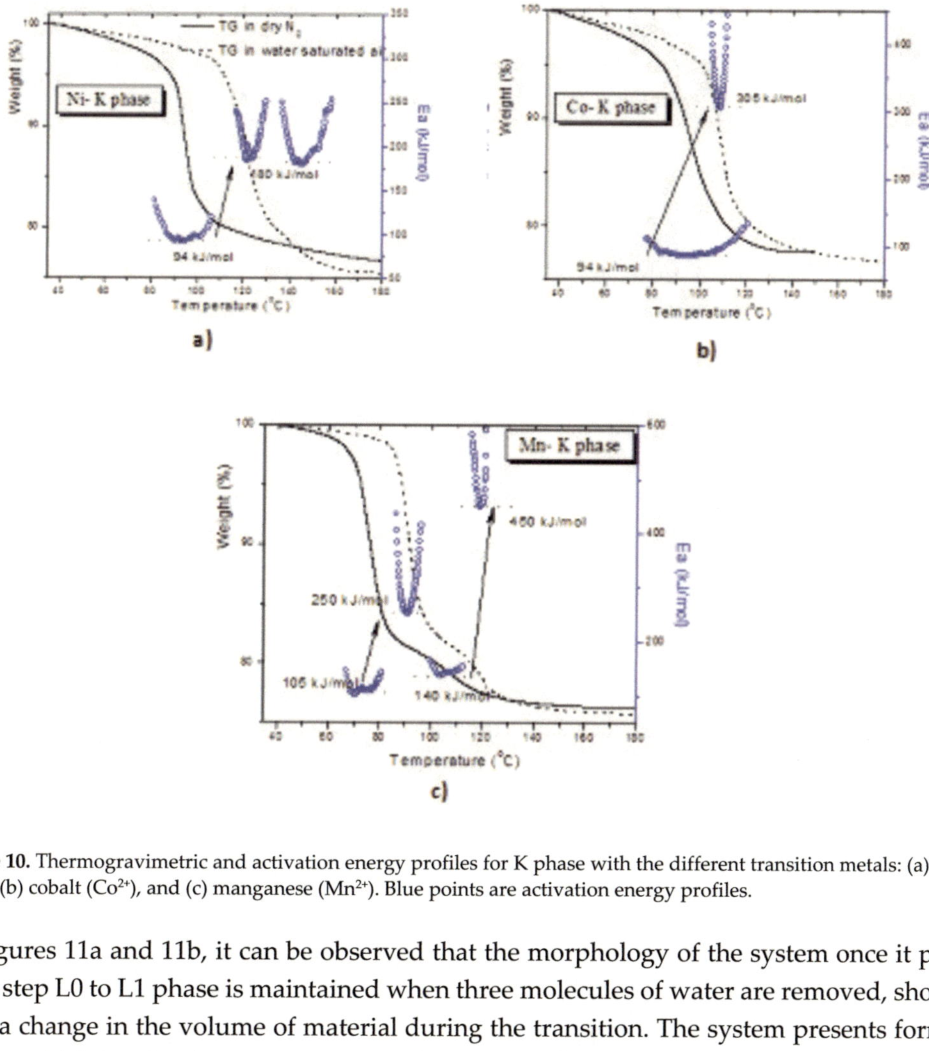

Figure 10. Thermogravimetric and activation energy profiles for K phase with the different transition metals: (a) Nickel (Ni^{2+}), (b) cobalt (Co^{2+}), and (c) manganese (Mn^{2+}). Blue points are activation energy profiles.

In Figures 11a and 11b, it can be observed that the morphology of the system once it passes from step L0 to L1 phase is maintained when three molecules of water are removed, showing only a change in the volume of material during the transition. The system presents forms as flakes, which is a laminar structure manifestation in these materials that crystallize. The L1 phase synthesized presents particle sizes smaller than that obtained by the transition.

In the characterization by SEM, the only sample that differs from the others is the phase L0 with nickel and external metal because this has very organized small flakes. Figure 12 shows that this structural arrangement explains why this material has low activation energies for the L1 phase derived from the corresponding L0.

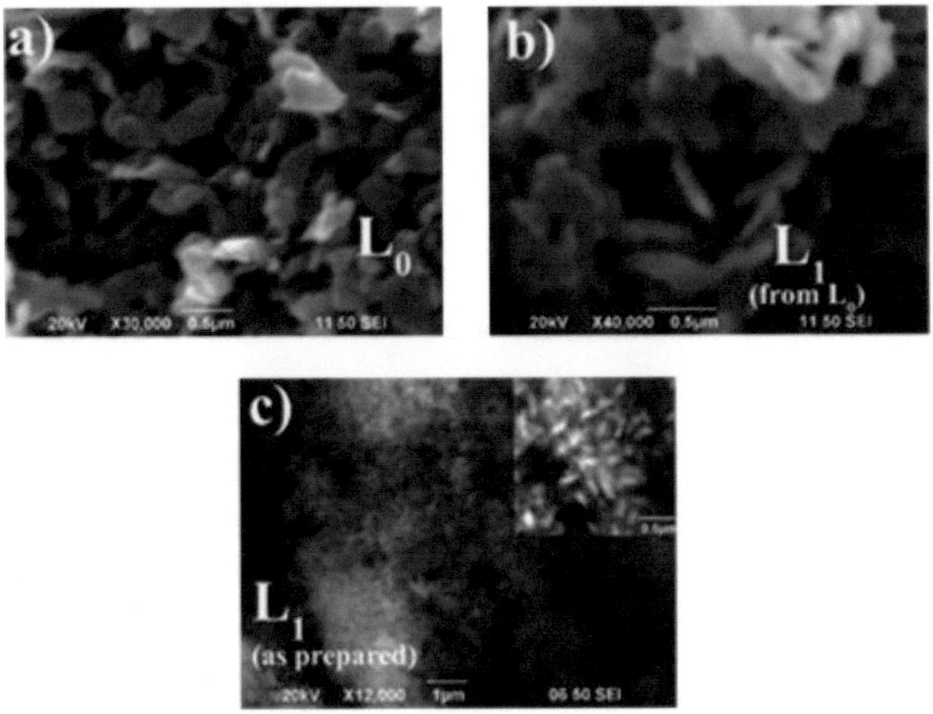

Figure 11. SEM micrographs of (a) as-prepared L0 phase, (b) L1 derived from L0 after heating, and (c) as-prepared L1 phase. The insets show a magnification of a selected area.

Figure 12. SEM micrograph of the as-prepared Ni-L0 phase.

2.7. Porous molecular materials

The second family studied is similar to the Prussian blue materials. These materials are referred to as hexacyano-cobaltates (M_3 [Co (CN) $_6$] $_{2x}H_2O$); M: Mn, Ni, Co, Zn, Cd and Cu (see Figure 13). These materials form an interesting family with windows and pore volumes appropriate for the separation and storage of small molecules [32]. The synthesized material boasts of water molecules in pores formed, and these can be reversibly moved through the window without having some structural rearrangement in the material. The unitary formula specifies that six water molecules are coordinated to the material, and the remaining space is occupied by water molecules that are linked weakly by hydrogen bonds mainly.

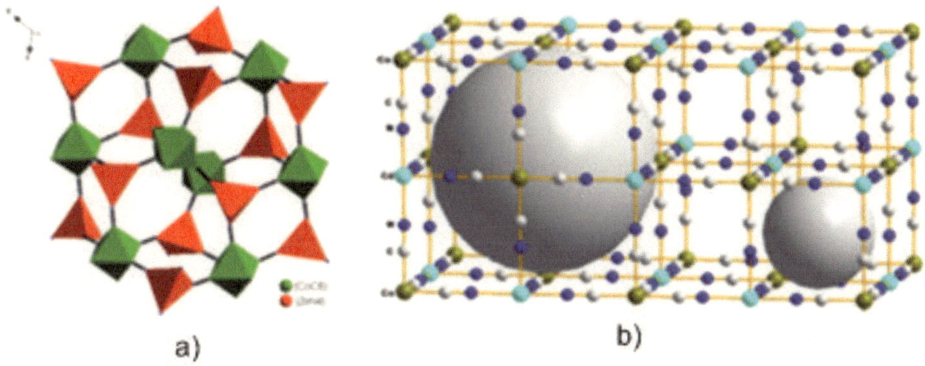

Figure 13. Lattice porous materials under study. (a) Representation of cavity Zinc cobalticyanide. (b) Porous lattice of the cubic phase in cadmium cobalticyanide.

The characterization in this family is carried out to evaluate the material according to its ability to remain at the same level and with the same crystal structure once the water molecules between the cavities are removed, as well as to obtain the corresponding activation energy profiles during dehydration as necessary energy values for the rupture of the interaction. Thus, this provides an idea of the power required for a molecule to be hosted inside the material.

2.7.1. Thermogravimetric molecular porous materials characterization

Materials under study lose the total number of molecules of water (both the coordinates and the zeolite) between 90°C and 180°C. The temperature dehydrated in a total way is highly linked with the polarizing power of the outer material, in addition to electrostatic forces with water molecules. Dehydration temperatures carry the following order: (117.54 °C) > Mn (106.57 °C) > Co (99.26 °C) > Cd (89.82 °C) > Cu> Zn (67.44 ° C). This sequence of dehydration can be seen clearly in Figure 14, in addition to showing the corresponding graphics of the derivative for each sample. In the particular case of the material with Zinc as metal coordination, low temperatures of dehydration due to the tendency of this material to adopt a tetrahedral coordination by forming a rhombohedral phase, which tends to be anhydrous. In these materials, for all the samples, the water molecules, both the coordinates and that of the zeolitic type, leave the system through a virtually continuous and simple process, resulting in dehydration cooperatively between the two types of water resulting in TG profiles without well-defined inflections, although at the beginning of all samples, a slight disturbance is recorded by the team (through the measurement of Ea).

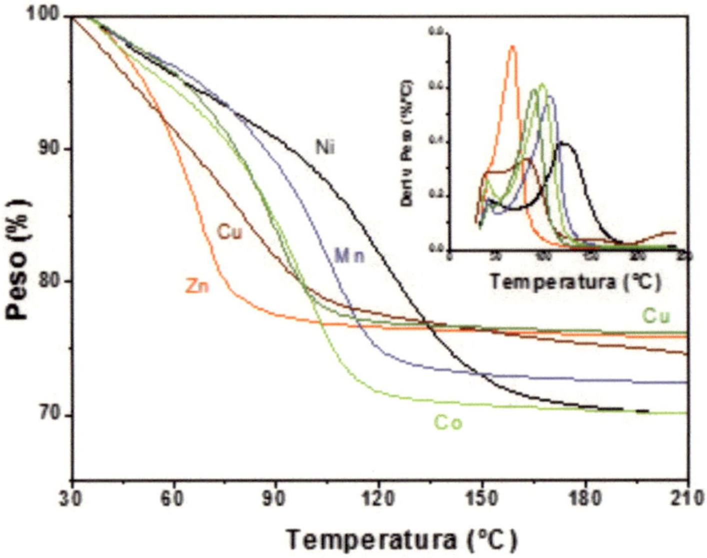

Figure 14. Thermogravimetric (TG) curves and the corresponding derivative (DTG) that expresses the progress of dehydration.

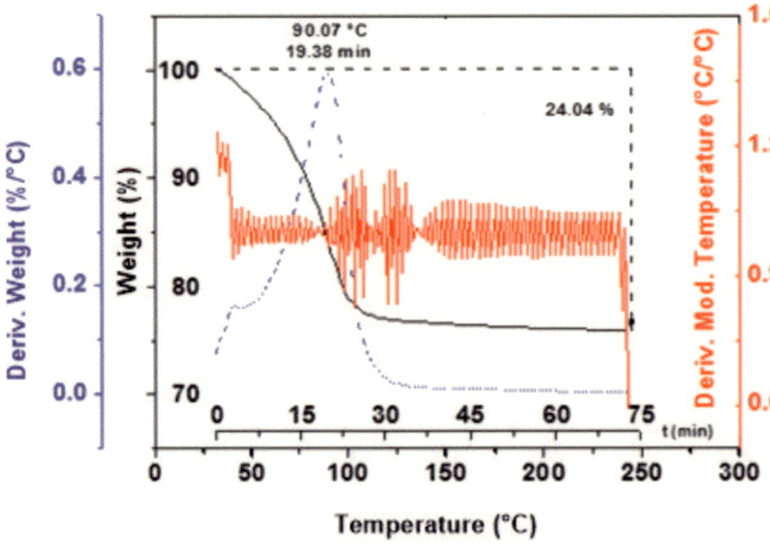

Figure 15. TG and DTG profile, modulation of temperature, and time set on a measure imposed on high resolution and modulated for material $Co_3\,[Co(CN)_6]_2xH_2O$.

http://dx.doi.org/10.5772/61076

As an example of a deeper analysis system, cadmium hexacyano-cobaltates are considered. In addition to being plotted and based on the weight loss derived from the corresponding weight, Figure 15 will also include the time of the process and the derivative of the temperature modulation time. The latter refers to the temperature disturbance along throughout the dehydration process, and the information provided is the resistance of the phenomenon, that is, as the modulation is increasing, the resistance of the water molecules evolves in the system. So, considering a constant disturbance implies that water molecules get out of the system without offering any more resistance. This agrees with what the Ea charts show.

2.7.2. *Hexacyano-cobaltates activation energy*

The activation energy profiles for these materials show that the Ea values are very close together, in a range between 60 and 90 kJ/mol with 60 kJ/mol in the case of Zinc as an external metal and 90 kJ/mol for nickel. Figure 16 shows the corresponding profiles mentioned. These two materials are considered because both ends are minimum and maximum in dehydration temperature.

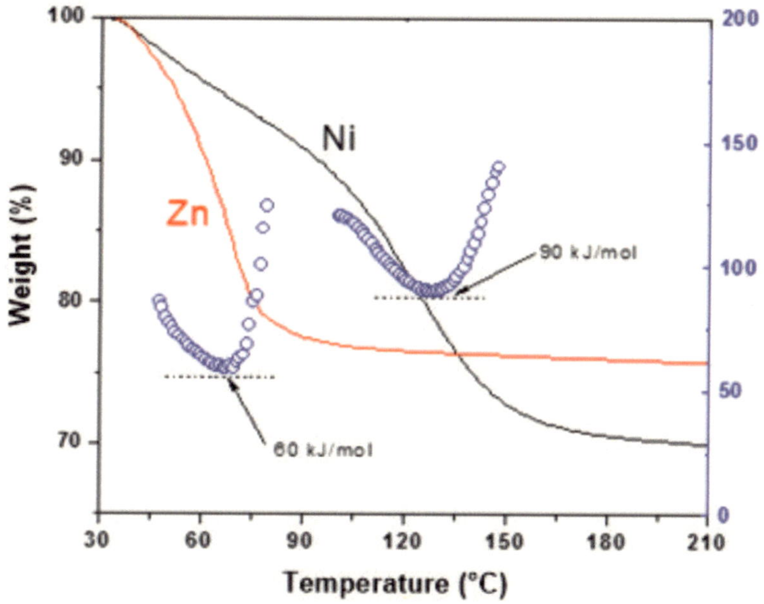

Figure 16. TG curves and profile of Ea for dewatering $M_3\,[Co(CN)_6]_2xH_2O$, con M = Ni^{2+} y Zn^{2+}.

3. Suggestions

In future research, we intend use a lattice Boltzmann method in order to copy the porous medium obtained experimentally and to make simulations. Figure 17 shows the simulation using the lattice Boltzmann method of randomly generated porous media [33].

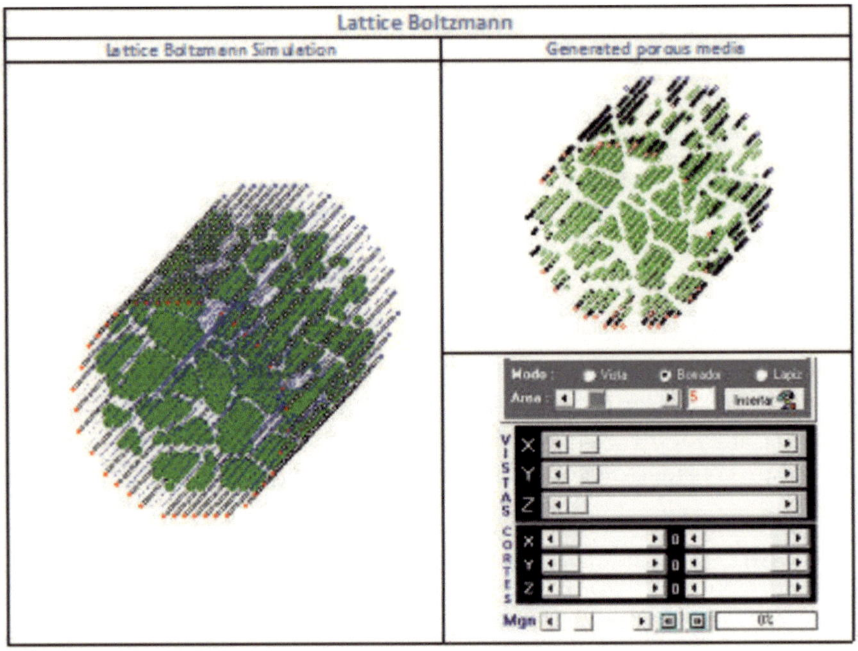

Figure 17. Lattice Boltzmann simulations through porous media.

4. Conclusions

In general, the study of the relative stability of the tetracyano-niquelates and hexacyano-cobaltates processes of dehydration shows that this depends in the first instance of the interactions of the water molecule with the outside. A more polarizing power metal is the stability of the link M-H_2O (for identical structures). Lamellar systems presented an energy barrier that must overcome the water molecules to begin to spread on the system. This barrier is the activation energy that lies in a range between 60 and 500 kJ/mol, depending on the type of water that comes out. In systems where the water molecules of subnetworks are homogeneous, water zeolite is indistinguishable from the coordinated water molecules.

To increase the partial pressure of water in laminar systems, there is an increase in the temperature of dehydration due to the increase in chemical potential of water in the system, and the immediate consequence is reflected in the increase in activation energy.

Studies by X-ray diffraction report the existence of a transition from L0 phase to L1 phase during the dehydration process, and during this transition, the phase change morphology of the material is maintained.

The process of dehydration in the hexacyano-cobaltates, which also presented the two types of water molecules (zeolite and coordinated), is carried out to lower activation energy values (between 60 and 90 kJ/mol) compared to lamellar systems since coordinated molecules are linked weakly to assemble metal.

For molecular materials during dehydration, there was no structural collapse, even if there are changes in the material locally, which has a shrinkage of only 4% of the unit cell.

In future research, we aim to use the lattice Boltzmann method for the characterization of porous medium, which would help with different configurations of porous media.

Author details

Omar Reyes-Martínez and José Luis Velázquez Ortega*

*Address all correspondence to: siulj@unam.mx

Facultad de Estudios Superiores Cuautitlán, UNAM, Cuautitlán Izcalli, Edo. de Méx, México

References

[1] Knoblauch, K.: Pressure swing adsorption geared for small volume users. Chem. Eng. 1978;85(25):87.

[2] Cheng C.D., Silvestri A.J.: The conversion of methanol and other O-compounds to hydrocarbons over zeolite catalysts. J. Catal. 1997;47(2);249–259. doi: 10.1016/0021-9517(77)90172-5.

[3] Schlapbach L., Züttel A.: Hydrogen-storage materials for mobile applications. Nature. 2001;414;353–358, doi:10.1038/35104634.

[4] Kitagawa S., Kitaura R.: Pillared layer compounds based on metal complexes. Synthesis and properties towards porous materials. Comments Mod. Chem. A. Comments Inorg. Chem. 2002;23(2);101–126. doi:10.1080/02603590214512.

[5] Mathey Y., Mazieres C.: Les phases cyanures de nickel(II) hydratés. Can. J. Chem. 1974;52(21):3637–3644. doi:10.1139/v74-544.

[6] Barrer R.M., MacLeod D.M.: Activation of montmorillonite by ion exchange and sorption complexes of tetra-alkyl ammonium montmorillonites. Trans. Faraday Soc. 1995;51:1290–1300. doi:10.1039/TF9555101290.

[7] Lemus-Santana A.A., Rodriguez-Hernandez J., del Castillo L.F., Basterrechea M., Reguera E.: Unique coordination of pyrazine in T [Ni(CN)4]·2pyz with T = Mn, Zn, Cd. J. Solid State Chem. 2009;182(4):757–766. doi:10.1016/j.jssc.2008.12.028.

[8] Rodriguez-Hernandez J., Lemus-Santana A.A., Cargas C.N., Reguera E.: Three structural modifications in the series of layered solids T(H2O)2 [Ni(CN)4]·xH2O with T =

Mn, Co, Ni: their nature and crystal structures. C. R. Chim. 2012;15(4):350–355. doi: 10.1016/j.crci.2011.11.004.

[9] Ferlay S., Mallah T., Ouahès R., Veillet P., Verdaguer M.: A room-temperature organometallic magnet based on Prussian blue. Nature. 1995:378;701–703. doi: 10.1038/378701a0.

[10] Shin-ichi O., Ken-ichi A., Yusuke S., Kazuhito H.: Humidity-induced magnetization and magnetic pole inversion in a cyano-bridged metal assembly. Nat. Mater. 2004;3:857–861. doi:10.1038/nmat1260.

[11] Lemus A.A.: Incorporación de pirazina y 4,4′-bipiridina en estructuras laminares de tetracianoniquelatos: síntesis, estructura, adsorción y separación de gases (Tesis), México, D.F., UNAM, 2010.

[12] Nijkamp M. G., Raaymakers J. E. M. J., van Dillen A. J., and de Jong K.P.: Hydrogen storage using physisorption – materials demands.Applied Physics A: Materials science and Processing; 2001; 72(5): 619-623. doi: 10.1007/s003390100847.

[13] Breck D.: Zeolite Molecular Sieves: Structure, Chemistry and Use. New York: J. Wiley; 1974. ISBN 0471099856, 9780471099857.

[14] Leonowicz M.E., Lawton J.A., Lawton S.L., Rubin M.K.: MCM-22: a molecular sieve with two independent multidimensional channel systems. Science. 1994;264(5167): 1910–1913. doi:10.1126/science.264.5167.1910.

[15] Thomas K.M.: Hydrogen adsorption and storage on porous materials. J. Catal. Today. 2007:120(3–4);389–398. doi:10.1016/j.cattod.2006.09.015.

[16] Zhao X.B., Xiao B., Fletcher A.J., Thomas K.M.: Hydrogen adsorption on functionalized nanoporous activated carbons. J. Phys. Chem. B.2005:109(18);8880–8888. doi: 10.1021/jp050080z.

[17] Zhao X., Villar R., Fletcher A.J., Thomas K.M.: Kinetic isotope effect for H2 and D2 quantum molecular sieving in adsorption/desorption on porous carbon materials. J. Phys. Chem. B.:2006;110(20):9947–9955. doi:10.1021/jp060748p.

[18] Panella B., Hirscher M., Roth S.: Hydrogen adsorption in different carbon nanostructures. Carbon, 2005:43(10):2209–2214. doi:10.1016/j.carbon.2005.03.037.

[19] Züttel A.,Sudan P., Mauron P., and Wenger P.: Model for the hydrogen adsorption on carbon nanostructures. Appl. Phys. A. Mater. Sci. Process. 2004;78(7);941–946, 2004. doi:10.1007/s00339-003-2412-1.

[20] Lachance P., Benard P.: Specific surface effects on the storage of hydrogen on carbon nanostructures. Int. J. Green Energy. 2007;4(4):377–384. doi: 10.1080/15435070701337434.

[21] Pinnavaia T.J.: Nanoporous layered materials. ACS Adv. Chem. Ser. 1995;245:283–300 doi:10.1021/ba-1995-0245.ix001.

[22] Bizeto M.A., Shiguihara A.L., Constantino V.R.L.: Layered niobate nanosheets: building blocks for advanced materials assembly. J. Mater. Chem. 2009;19(17):2512–2525. doi:10.1039/B821435B.

[23] Vieira P. and Santos C.A.C.: Propriedades catalíticas dos argilominerais (Parte II): Argilominerais intercalados e pilarizados. Boletín Técnico de Petrobras. 1988;31(2):143–157.

[24] Sterte J.: Hydrothermal treatment of hydroxycation precursor solutions. Catal. Today. 1988;2(2):219–231. doi:10.1016/0920-5861(88)85005-3.

[25] Dunbar K.R., Heintz R.A.: Chemistry of transition metal cyanide compounds: modern perspectives. Prog. Inorg. Chem. 1997;45:295–296. doi:10.1002/9780470166468.ch4.

[26] Iwamoto T., Nishikiori S.-I., Kitazawa Y.H.: Dalton transactions. J. Chem. Soc. 1997;22:4127–4136. doi:10.1039/A702539D.

[27] Ludi A.: Prussian blue, an inorganic evergreen. J. Chem. Educ. 1981;58:1013. doi: 10.1021/ed058p1013.

[28] Ludi A., Gudel H.U.: Structural chemistry of polynuclear transition metal cyanides. J. Inorg. Chem. 1973;14:1–21 doi:10.1007/BFb0016869.

[29] Roque J., Reguera E., Balmaseda J., Rodriguez H.J., Reguera L., del castillo L.F.: Porous hexacyanocobaltates(III): role of the metal on the framework properties. Microporous Mesoporous Mater. 2007;103:57–71. doi:10.1016/j.micromeso.2007.01.030.

[30] Kaye S.S., Long J.R.: Hydrogen storage in the dehydrated Prussian blue analogues M3 [Co(CN)6]2 (M = Mn, Fe, Co, Ni, Cu, Zn). J. Am. Chem. Soc. 2005;127:6506–6507. doi:10.1021/ja051168t.

[31] Chapman K.W., Southon P.D.,Weeks C.L., Kepert C.J.: Reversible hydrogen gas uptake in nanoporous Prussian Blue analogues. J. Chem. Commun. 2005;26:3322–3324. doi:10.1039/B502850G

[32] Boxhoorn G., Moolhuysen J., Coolegem J.G.F., van Santen S.R.: Cyanometallates: an underestimated class of molecular sieves. J. Chem. Soc., Chem. Commun. 1985;19:1305. doi:10.1039/C39850001305.

[33] Romo S.R., Ortega J.L.V.Simulation of non-Newtonian fluids through porous media. Chapter 5. In Patel V (ed.),Petrochemicals.Rijeka, Croatia: Intech; 2012:75–100. doi: 10.5772/2069. ISBN 978-953-51-0411-7.

Chapter 4

Crude Oil Desalting Process

Juan Pereira, Ingrid Velasquez, Ronald Blanco, Meraldo Sanchez,
César Pernalete and Carlos Canelón

Additional information is available at the end of the chapter

http://dx.doi.org/10.5772/61274

Abstract

Desalting is a water-washing operation performed initially at the production field and thereafter at the refinery site for additional crude oil cleanup. Salt and water content specifications are even more rigid because of their negative effect in downstream processes due to corrosion, and catalyst deactivation. An optimum formulation concept is presented to describe emulsion breaking in desalting process. In the stabilization mechanism is accepted that water droplets are stabilized by the formation of a mechanically strong and viscoelastic film at the interface composed of asphaltenes. In the case of water-in-crude-oil emulsions, a balanced optimum formulation is attained by adding to the lipophilic natural surfactants contained in the crude oil, demulsifiers which are hydrophilic. The aim is to relate the nature and concentration of the added demulsifier products to the amphiphilic mixture at the interface. All formulation parameters, such as solvent, alcohols, kind and concentration of demulsifier, amoung others, can be explained for proportional and saturation regimens.

Keywords: Desalting, asphaltenes, demulsifier, , emulsions, physicochemical formulation

1. Introduction

Petroleum recovered from reservoir is mixed with a variety of substances such as gases, water, chloride salts, and dirt, which contain other minerals. Thus, petroleum processing actually commences shortly after the production of fluids from reservoir, where pretreatment opera-

INTECH
open science | open minds

© 2015 The Author(s). Licensee InTech. This chapter is distributed under the terms of the Creative Commons Attribution License (http://creativecommons.org/licenses/by/3.0), which permits unrestricted use, distribution, and reproduction in any medium, provided the original work is properly cited.

tions are applied to the crude oil prior to transportation. Any crude oil to be shipped by pipeline, or by any other form of transportation, must meet strict regulations in regard to water and salt content.

Desalting is a water-washing operation performed initially at the production field and thereafter at the refinery site for additional crude oil cleanup, where the salt and water content specifications are even more rigid because of their negative effect in the downstream processes due to scale formation, corrosion, and catalyst deactivation. Desalting involves mixing heated crude oil with washing water, using a mixing valve or static mixers to ensure a proper contact between the crude oil and the water, and then passing it to a separating vessel, where a proper separation between the aqueous and organic phases is achieved.

Since emulsions can be formed in this process, there is a risk of water carryover in the organic phase. In order to overcome this problem chemical demulsifiers are added to promote the emulsion breaking. When this operation is performed at a refinery, an electric field across the settling vessel is applied to coalesce the polar salty water droplets, and, therefore, a decreasing in water and salt content is achieved.

The understanding of the different variables that affect the desalting process, especially the effect of the amount of chemical demulsifiers used, is imperative in order to optimize operating costs. According to data published by Vafajoo [1], the demulsifier concentration used for desalting could reach up to 100 ppm. Considering a demulsifier cost about 1.5 $/lb [2] for a medium capacity refinery (200 MBPD), the operating cost related to chemical demulsifiers would be in the order of 2 MM$/year. Thus, any reduction in demulsifier concentration in crude oil desalting would generate savings in the order of many thousand dollars. In the current work, a technical description of desalting crude oil and a wide description of stability emulsion phenomena are carried out. Finally, this work aims mainly at understanding the demulsifiers role from the viewpoint of physicochemical formulation.

2. Process description

Salts in crude oil are mainly in the form of magnesium, calcium, and sodium chlorides, sodium chloride being the most abundant. These salts can be found in two forms: dissolved in emulsified water droplets in the crude oil, as a water-in-oil emulsion, or crystallized and suspended solids.

The negative effect of these salts in downstream processes can be summarized as follows: salt deposit formation as scales where water-to-steam phase change takes place and corrosion by hydrochloric acid formation. Hydrochloric acid is formed by magnesium and calcium chlorides' decomposition at high temperatures (about 350 °C) as follows [3]:

$$CaCl_2 + 2H_2O \rightarrow Ca(OH)_2 + 2HCl$$

$$MgCl_2 + 2H_2O \rightarrow Mg(OH)_2 + 2HCl$$

In addition, other metals in inorganic compounds present in reservoir dirt and sand produce catalyst poisoning in downstream processes such as hydrotreaters and cat crackers because of they are chemically adsorbed on the catalyst surface.

The objective of desalting process is to remove chloride salts and other minerals from the crude oil by water-washing. Depending on the desired salt content in the desalted crude oil, a one- or two-step process could be applied. For refining purposes, a salt concentration of maximum 1, 5 PTB (pound of salt measured as NaCl per thousand barrels) is desired. By desalting, a considerable percentage of suspended solids (sand, clay, or soil particles, or even particles product from corrosion of pipelines and other upstream equipments) are removed. Figure 1 shows a general process flow diagram for one- and two-step desalting process.

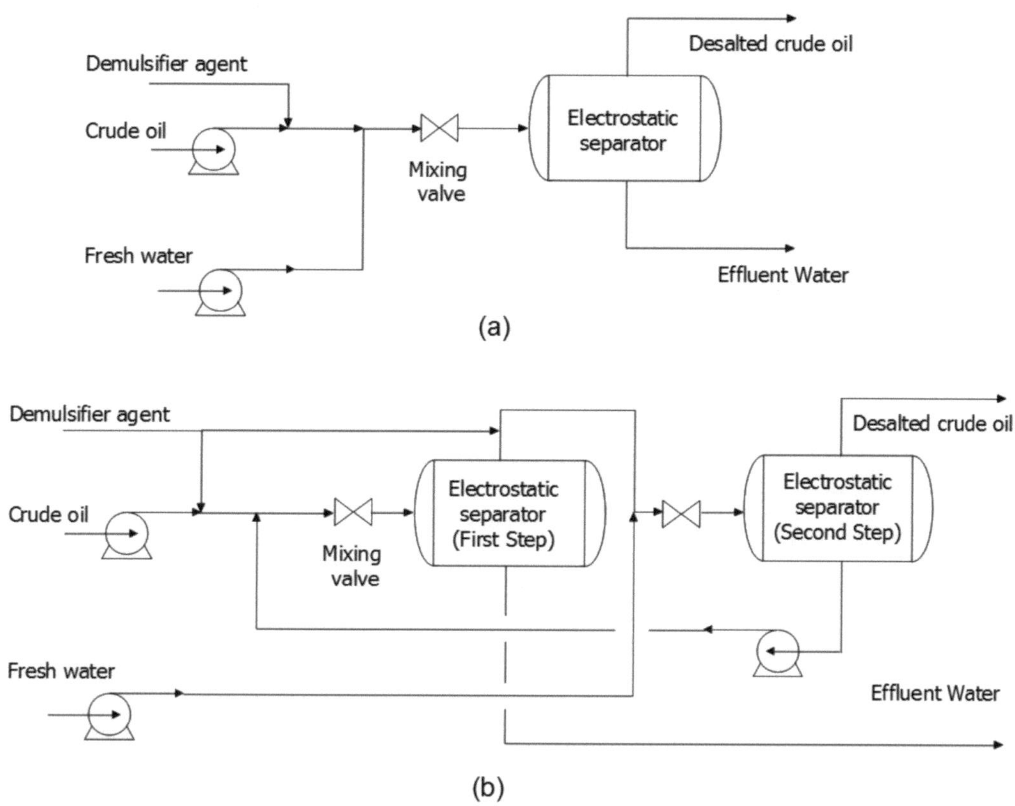

Figure 1. Simplified desalting process flow diagrams for different configurations: (a) one-step and (b) two-step.

The process starts by mixing the raw crude oil with an appropriate amount of washing water also known as dilution water. The washing water as a volume percent of the crude oil processed could oscillate between 3 and 10%, depending on the API gravity of the crude oil – the heavier the crude oil, the more the water required [2]. Demulsifiers are added to the crude oil in this process step.

In order to promote the effective mixing between the organic and aqueous phases, and ensure the proper dilution of the salts and minerals in the aqueous phase, a mixing valve is used. This is a common globe valvewhich causes a pressure drop and, as a result, a shear stress over the droplets that promotes an intimate water and oil mixture. The main aspect that needs to be considered is the pressure drop, whose value it is about 10-50 psi and varies according to the flux through the valve (automatic differential pressure controllers could be used). In addition to the mixing valve, upstream premixing device could be used, such as spray nozzles at the point of water injection or static mixers, between the water injection point and the mixing valve [4]. High delta pressure in the mixing valve promotes smaller droplets, which is positive because it improves the contact among the phases; however, very small droplets could yield a more stable emulsion, which could cause problems in the separating vessel. Therefore, it is very important to balance both effects in the selection of the operation pressure drop.

Then this mixture goes to the desalter, a horizontal cylindrical tank that provides long enough residence time to separate the water and oil mixture in two phases. Some water droplets diameters are so small that they could not be separated by gravity; so, an electrostatic field between two electrodes installed into the desalter is used to promote coalescence. Due to the dipolar nature of the water molecule, this electric field promotes an attraction with the other water molecules in the neighbor droplets promoting coalescence. According to Gary et Al. [2], either AC or DC fields may be used and potentials from 12, 000 to 35, 000 volts. The attraction force (*F*) between the water droplets is given by:

$$F = \frac{K_s \cdot \varepsilon^2 \cdot d^6}{s^4} \tag{1}$$

where ε is voltage gradient, d is droplet diameter, s is the distance between drops centers and K_s is a constant for the system [4].

Finally, after coalescence, water droplets settle according to the well-known Stock's law given by:

$$Settling\ rate = \frac{k \cdot \left(\rho_{water} - \rho_{oil}\right) \cdot d^2}{\mu_{oil}} \tag{2}$$

where k is a constant, d is the droplet diameter, ρ is density, and μ is viscosity.

In a one-step configuration, shown in Figure 1(a), 90% of salt removal can be achieved. For higher salt removal percentages, a two-step configuration, shown in Figure 1(b), should be required. As far as the last process configuration is concerned two demulsifier injection points are used, both before the mixing valve in the first and second step. In addition, freshwater is fed to the second stage, and effluent water from this stage is recycled to the first one. With this configuration, a 99% salt removal could be achieved.

3. Crude oil emulsions

Emulsions are present in the main operations of the petroleum industry. The more well-known emulsions in the petroleum industry are those when the crude oil is the external phase (W/O). Water-in-oil emulsions are robustly stabilized by natural surfactants from crude oil, where a "special film" takes place and it is the responsible for emulsion stability. Film at the oil–water interface is highly viscous and is formed by asphaltenes, resins, waxes, and small solid particles [5, 6].

A main feature for interfacial film is the irreversible adsorption of asphaltenes at the oil–water interface in combination with other natural surfactants [7-10].

In the study of water-in-oil emulsions, it is important to know the structure and properties of the crude oil components, their tendency to associate and accumulate at the interface, as well as solubility and sensitivity to pressure and temperature changes.

3.1. Formation of water-in-oil emulsions

In secondary oil recovery processes, the oil is contacted with formation water or water injected to produce the oil drain water. Water is used as a piston to push the oil from the well during the production process and then remove to the refinery [11]. In the reservoir, the fluid velocity is very slow (1 ft/day) to produce the emulsion, and emulsion is not formed during the two phase flow in porous media but later in equipment where oil is processed. Hence, the shear is responsible of making emulsion by pumping through valves, pipes, elbows, and others [7].

Damage to environment may occurs when petroleum or its products are spilled into the sea or the river, due to which water–oil emulsions are also formed. These emulsions (called "chocolate mousse") has special properties and characteristics [12-13]. Changes in the emulsions properties, such as viscosity and stability, are notable due to internal phase increasing [13].

The formation of water/oil emulsions is generally caused by the presence of resins and asphaltenes present in the oil, which play the role of natural emulsifiers [9]. These emulsifiers have a mutual attraction resulting in the formation of an elastic membrane around the droplets, preventing water droplets from uniting and decanting by gravity. This membrane is thick and could be easily visible using an optical microscope [14].

3.2. Asphaltenes

Asphaltene is the petroleum most complex fraction which is insoluble in normal alkanes such as n-pentane, n-hexane, or n-heptane, but soluble in benzene or toluene. Asphaltenes are composed by high molecular weight molecules and the most polar components in crude oil [15].

Usually, asphaltene structure comprises fused polyaromatic rings substituted by short aliphatic chains (1-3 carbons atoms) and cycloalkanes. Aromatic rings containing some polar functional group than content heteroatom such as sulfur, nitrogen and oxygen and some metals like nickel, vanadium and iron, which confer polarity and amphiphilic characteristics to these macromolecules [16]. Asphaltenes can generate a wide distribution of molecular structures that can vary greatly from one crude to another [17].

Asphaltenes are composed of several polynuclear aromatic sheets surrounded by hydrocarbon tails, and form particles whose molar masses are between 500 and 20.000 g. They contain many functional groups, including some acids and bases [18].

The specific structure of asphaltenes is unknown; however, molecular weight averages about 750 and there is a planar aromatic structure surrounded by alkane groups, some with heteroatoms, sulfur (S), nitrogen (N), and oxygen (O) [19]. Deposition of asphaltenes often causes some changes in the flow behavior of the oil fields through processing facilities, and difficulties in oil recovery. The problems caused by asphaltenes are closely related to its stability. Due to pressure variations, oil composition, and temperature, asphaltenes tend to form aggregates [14].

Experimental evidence from a combination of spectroscopy techniques and elemental analysis suggest that resins contains several fused five-and-six carbons rings in addition to alkyl chains.

Crude oil can be defined as a dispersed system where asphaltenes are colloids. Asphaltene molecules are in the center of the structure of the liquid oil, the other components are organized around the layer of asphaltenes decreasing polarity (resins, saturated and aromatic hydrocarbons). This simple model allows reconciling the compatibility presence within the same fluid and asphaltenes.

3.3. Mechanism of asphaltene aggregation: Colloidal behavior

Asphaltenes can be considered as a complex polydisperse system of molecules in which subcomponents with significantly different characteristics can be distinguished as the result of solubility differences [17, 20].

From the colloidal point of view, it has been stated that asphaltenes in general are a blend of two main fractions. Acevedo et al. [21] proposed a model for asphaltene colloids in toluene which consists of two subfractions with different solubilities, called A1 and A2. The subfraction A1 (90 mg/L in toluene), which is insoluble in aromatic solvents such as toluene, cumene, and others, is found in greater proportion than a second subfraction A2 (57 g/L in toluene), which is soluble in these solvents and acts as a promotor in the dispersion of A1. The solubility of the latter is similar to the total mixture. Asphaltene aggregation phenomena is well understanding under the vision of colloidal nature of both fractions. [22]. Furthermore, this colloidal model has been found very useful in accounting for other phenomena such as trapping of compounds, flocculation, aggregation promoted by A1, asphaltene aggregation at very low concentration, and metal distribution [17].

In the vapor pressure osmometry (VPO) technique for molecular weight measures, the polarity of solvent is a critical parameter. Using o-dichlorobenzene, a 1000 g mol^{-1} value was reported for A2 solutions at 120 °C. On the other side, A1 afforded higher values (around 2600 g mol^{-1}), emphasizing the strong aggregation tendency of this fraction [23].

Asphaltenes can form different types of aggregates depending on the environment: oil, solvent, interface, and others. Mullins et al. [24] proposed a scheme to explain the aggregation of asphaltenes in toluene. At low concentration, asphaltene is a group of individual molecules that evolve into nanoclusters and are associated in particles until reaching a system susceptible to flocculate. Floccules are unstable aggregates that eventually precipitate. In certain cases, stable aggregates that give the crude viscoelastic properties are formed.

Due to the presence of polar groups, asphaltenes can behave as surfactants, exhibiting the phenomena of adsorption and aggregation. However, it is wrong to attribute a critical micelle concentration (CMC) for asphaltenes and other macromolecules from crude oil. CMC is a parameter exclusive for surfactants, such as sodium dodecyl sulfate ($8.3 * 10^{-3}$ M in water). Parameters such as aggregation number, shape and size of the micelle, electrolyte effect on the micellization and other additives are well known for surfactants. For petroleum asphaltenes this knowledge does not exist, mainly because the structure of an asphaltenes molecule is unknown. Aggregates phenomena best describes asphaltenes behavior. Asphaltene in toluene begins to form aggregates at 50 ppm concentration [25].

In summary, the use of CMC term for asphaltenes and other natural surfactants contained in the crude oil should be avoided. The emulsions in crude oil are produced for a mechanism starting with asphaltenes migrate into the oil–water interface, a process which is regulated by the diffusion of the soluble asphaltenes [19]. Adsorption of asphaltenes at the interfaces is a very slow process. The initial diffusion step toward the interface is rapid and it is followed by a long reorganization and the progressive building of multilayers. Asphaltenes diffuse from the bulk of crude oil to oil–water interface and are then adsorbed, decreasing the interfacial tension [26]. Asphaltenes can also form multilayers on solid surfaces such as silica. It has been found that increasing the concentration of asphaltenes in solution tends to increase its state of aggregation, which decreases the rate of adsorption on the surface [22].

3.4. Stability of water-in-crude-oil emulsions

In terms of time, the stability of the emulsion has a range from few minutes to years and it varies by the characteristics of the crude oil and water [27]. The stability mechanism is, of course, related to the composition of crude oil, such as type and the concentration of natural surfactants, the physical properties of oil, and the interfacial rheology of the interface around water droplet, which provides information about elasticity and viscosity and, for instance, the presence of a film (called skin) for the specific case of petroleum crude oil at this interface [10, 28].

Gao et al. [29] quantitatively studied the formation of interfacial film by the asphaltenes on the oil–water interface. They propose to measure the appearance of crumpling in the interfacial film, using imaging drop following compression by extraction volume, by the following relationship:

$$CR = \frac{A_f}{A_i} = \frac{\pi R_{f^2}}{\pi R_{i^2}} = \frac{R_{f^2}}{R_{i^2}} \tag{3}$$

where CR is crumpling ratio; A_i is the projected drop initial area; A_f is the projected area of the drop observed just before crumpling; R_i is the projected drop initial radio; and R_f is the radio of the drop observed just before crumpling. Figure 2 shows a schema of the contraction of a drop of oil in water. In Figure 2 (a), crumpling phenomenon does not occur; drop reduces its volume to be sucked into the syringe. In Figure 2 (b), volume drop is contracted when the drop crumpling phenomenon occurs in the interfacial film. The scheme shows its building. CR parameter is related to the irreversible adsorption of asphaltenes at the water–crude interface; an increasing of CR means an increasing in emulsion stability.

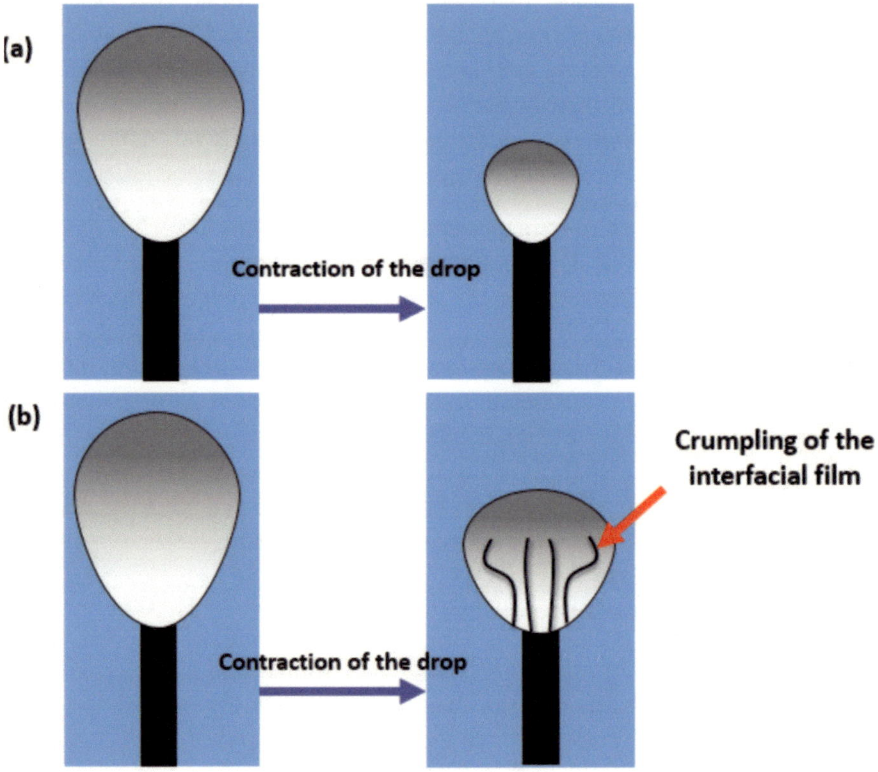

Figure 2. Volume contraction scheme of a drop of oil in water system: (a) absence of crumpling phenomenon and (b) presence of crumpling phenomenon occurs.

As part of the stabilization mechanism, it is accepted that the water droplets are stabilized by the formation of a mechanically strong and viscoelastic film at the interface, composed of asphaltenes [30].

Subsequently, several workers reviewed emulsions and concluded that asphaltene content is the most important factor in the formation of emulsions [7-10]. Asphaltene fraction is mainly responsible for the high emulsion stability of water-in-crude-oil; thus, they were found to be capable of forming rigid, cross-linked, and elastic films [19].

Asphaltenes stabilize water-in-oil emulsions if they are near or above the point of incipient flocculation, i.e., solid particles which can be. Other investigators have also suggested that asphaltene colloids are responsible for stable emulsions. Asphaltenes may be at the interface as fine solid particles or resin-asphaltene colloids. However, Yarranton et al. [31] reported that, at low asphaltene concentrations (<0.2 wt %), molecular state of asphaltenes predominant onto the colloidal form. It is also necessary to consider the effect of resins as they may influence asphaltene aggregation, adsorption on the interface, and emulsion stability [32].

One of the most used methods for studying emulsion stability is the bottle test. Figure 3 shows the bottle test sample for evaluating the stability of water-in-oil emulsion. It was observed that in the tubes on the left side there are stable emulsions, while in the tubes in the center and on the right side, there is a good separation of the two phases.

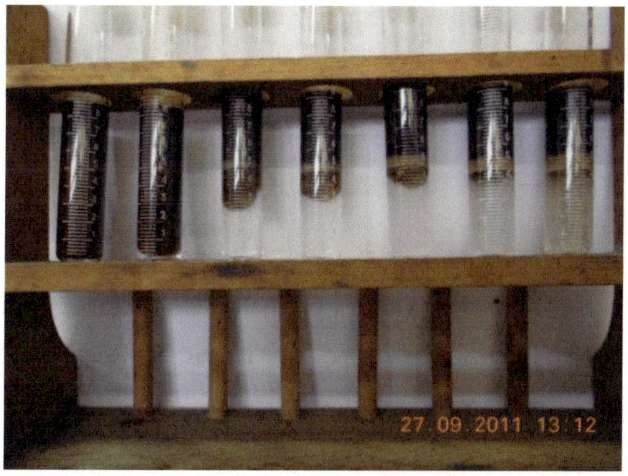

Figure 3. Stability bottle test of water-in-crude-oil emulsion (Silva, 2012).

Langmuir technique has been widely used for the characterization of the interfacial film formed by natural surfactants from crude oil [33-37]. Isotherms provide surface-pressure-area information on the compressibility of the film formed by the asphaltenes and other natural surfactants. The interfacial film can be collected with the Langmuir-Blodgett technique. Extensive morphological characterization of the interfacial film has been performed, including properties such as contact angle, thickness, and aggregate size, among other. Images obtained using Atomic Force Microscopy have contributed to the understanding of the mechanisms of interfacial adsorption, swelling, aggregation of colloidal interactions, and asphaltene [38, 39].

Aromaticity, length of the alkyl side chains, and branch reduction in the aliphatic chains of resins and asphaltenes are associated with the formation of stable emulsions. Another important parameter is the pH of the aqueous phase. Asphaltenes and resins are composed of polar functional groups; these groups can be ionized at alkaline pH. Surface charge density increases due to ionization of polar groups, generating internal repulsive forces destroying the mechanical properties preventing droplet coalescence [14].

4. Chemical demulsification: physicochemical formulation

When emulsion formation and stabilization take place, demulsifiers (generally a surfactant) must be used. A surfactant is a specie that, when present at low concentration in a medium, has the property of adsorbing onto the surfaces or interfaces of the system, altering the surface or interfacial free energies of those surfaces (or interfaces) [40].

Balsamo et al. [41] showed that effective demulsifiers formed complexes with natural anionic surfactants present in the oil phase, weakening the irreversible film; this is well correlated with low emulsion stability. When emulsion breaker was added to the system, a synergistic effect was observed between natural surfactant and demulsifier with further lowering of the interfacial tension (IFT).

It is known that an unstable emulsion occurs when amphiphilic formulation at the interface corresponds to an exact affinity balance of the surfactant mixture for the oil and water phase [7, 42-44]. From a physicochemical viewpoint, the optimum formulation is achieved when the standard potentials (μ^*) of a surfactant (or mixture) in oil (μ_o^*) and water (μ_w^*) are equal, i.e., when (μ_w^*)=(μ_o^*). Physicochemical phenomena have been qualitatively related to the emulsion breaking occurrence. In the case of water/crude emulsions, the balanced optimum formulation is attained by adding hydrophilic demulsifiers to the lipophilic natural surfactants contained in the crude oil. From our point of view the aim is relate to the nature and concentration of the added demulsifier products to the amphiphilic mixture at the interface.

A new methodology has been presented by several authors [7, 42-44]. It is essentially a formulation scan study carried out by adding a hydrophilic demulsifier at different concentrations, until a minimum emulsion stability is reached. At this optimum formulation point, the lipophilic natural surfactants (i.e., asphaltenes, resins, naphthenic acids, and other crude oil amphiphiles at the interface) referred to as asphaltenes (A) are balanced by the added hydrophilic demulsifiers (D). The hydrophilic–lipophilic contributions of both types in the interfacial mixture are dependent on their adsorbed concentration and on the strength of their effects. Consequently, the hydrophilic contribution of the demulsifier may be varied by changing its hydrophilicity or its concentration [44]. In general, it has been shown that the more hydrophilic the nature of the demulsifier, the lower is the concentration required to reach the minimum emulsion stability when they join asphaltenes [44], mixing rule is approximately linear, i.e., that the demulsifier is not too hydrophilic (because it would go essentially to water) or not too close to matching hydrophobicity (because it would not compensate the asphaltene hydrophobic effect). However, there are others factors to be considered such as the viscoelastic properties of the asphaltene interfacial layer than improve the understanding of demulsifier action.

Pensini et al. [45] measured the viscoelastic properties of the interface as a function of aging time as the asphaltenes diffused toward the interface and adsorbed on the water–toluene interface. With gradual accumulation of asphaltenes on the toluene–water interface, the interface changes from a viscous to an elastic-dominant microstructure. Focusing on the effect of demulsifier addition, at 6000 s of interfacial aging are not shown. Once the condition G′

http://dx.doi.org/10.5772/61274

(elastic component) = G″ (viscous component) was reached, the demulsifier was added at 2.3 ppm (optimum dosage from the bottle tests). When the elastic contribution is predominant, asphaltenes are intimately linked, forming a cohesive interfacial network. To assess the demulsifier performance, in penetrating and breaking the cohesive network of asphaltenes, the shear viscoelastic moduli was measured. Upon addition of 2.3 ppm demulsifier in toluene, both viscous and elastic shear modulus of the interfacial film decreased.

The problem when handling formulation issues in a crude oil system is that only a small part of the asphaltenes is present at the interface. This implies that the total asphaltenes' concentration in the crude oil does not participate with the demulsifier at the interfacial mixture. As a consequence, the optimum concentration of the demulsifier is not dependent on its mixture with all the asphaltenes in the system. Only a fraction of asphaltenes is adsorbed, and the rest forms aggregates, flocculates, or precipitates, an important amount of it being segregated close to the interface, but not at the interface (see Figure 4). This asphaltene excess is responsible for the characteristic skin at the water–crude oil interface [7, 9, 10].

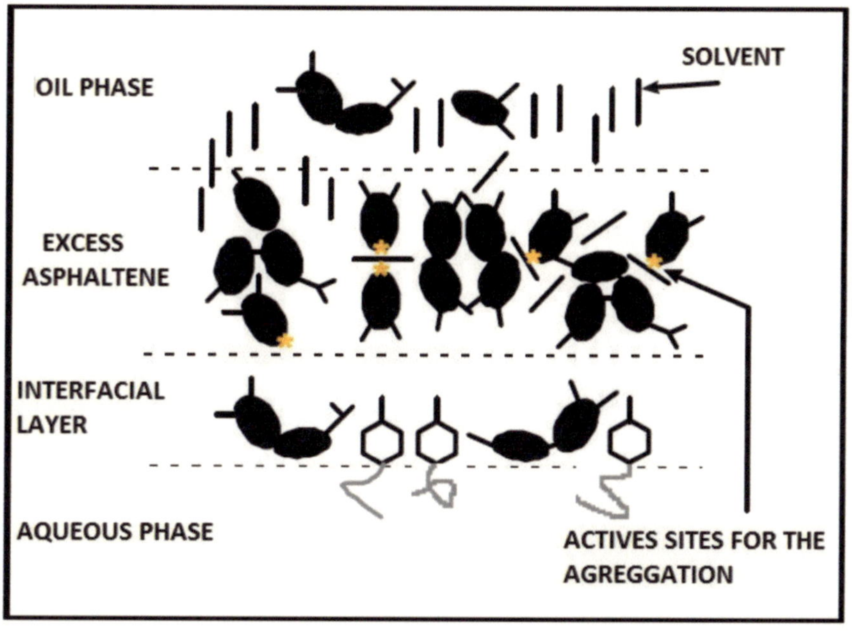

Figure 4. Asphaltene excess than there are out of interface. This situation corresponds to Ca> 1, 000 ppm. Solvent can to penetrate in this layer asphaltene and interact with them for limit its aggregation due solvation effects (Pereira, 2009).

An experimental method has been recently proposed to estimate the asphaltene–demulsifier adsorbed mixture composition, so that interfacial formulation may be estimated [7]. It consists in diluting the crude oil so that the asphaltenes' concentration in the oil phase is reduced. At some point, asphaltenes' concentration is low enough so there is no excess of aggregates or flocculates in the oil phase. In this situation, asphaltenes and demulsifiers species may be considered as two types of surfactants adsorbed at the interface, according to the physicochemical conditions in the bulk phases at equilibrium. It has been found that below a concentration of asphaltenes 500–1000 ppm, the concentration of demulsifier required to attain optimum formulation varies proportionally to the concentration of asphaltenes (see Figure 5). In this so-called proportional regime [7], a quantitative formulation effect at the interface can be estimated without taking into account physical effects such as viscosity, convection, and diffusion through bulk oil (these are not formulation issues). This method allows evaluating the demulsification performance as a function of different formulation variables (demulsifier hydrophilicity, salinity, and crude oil characteristics) [7, 42-44], which represents a significant practical progress in designing commercial products.

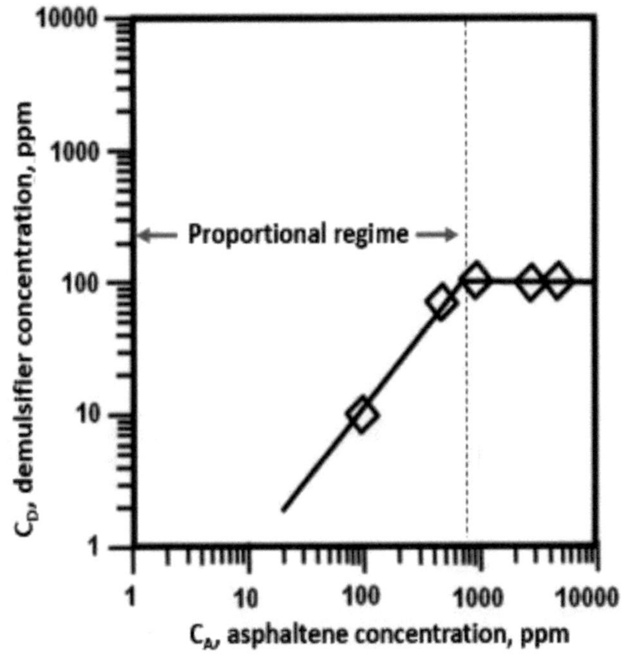

Figure 5. Demulsifier concentration as a function of the asphaltene concentration to reach optimum formulation (Pereira, 2009).

HLB method (hydrophile–lipophile balance) is often used. In this method, a number (0–40), indicative of emulsification behavior and related to the balance between the hydrophilic and lipophilic (hydrophobic) portions of the molecule, has been assigned to many commercial surfactants [42].

Using HLB concept, the proportional regime can be estimated, as follows, [7, 42-44]:

$$HLB_{optMixt} = X_A HLB_A + X_D HLB_D \tag{4}$$

where X_A and X_D are the fractions of A and D at the interface ($X_D = 1 - X_A$) and $HLB_{OptMixt}$ is a characteristic value (close to 10 correspond to the affinity) for the system. This can be calculated from the experimental correlation (bottle test) when the crude, brine, and temperature are fixed. The effect of any additive can be easily qualified as favorable or not when decreasing the fraction of demulsifier required (X_D) to attain the optimum formulation for the lowest stability. Since the experiments are much easier and more accurate in the proportional regime, this is the zone where the tests of the additive effects are mostly carried out. Silva et al. [8] used this methodology to estimate, in a semiquantitative manner, the effects of organic acid and amine additives on the formulation of the components at the interface. The results indicate that water-soluble additives (acetic acid and ethylamine) have no significant effect on the required demulsifier concentration C_D*. Even if they considerably alter the pH aqueous phase, additives do not reach a value to ionize asphaltenes species. The additives of lipophilic nature (hexanoic acid and hexylamine) both interact on the other side of the interface (i.e., asphaltenes in the oil phase). The hexanoic acid is found to marginally increase the emulsion stability, probably because it could lead to a slightly more hydrophobic co-surfactant effect. In contrast, the hexylamine, which is likely to interact with the asphaltenes through an acid–base mechanism, produces the delocalization of charges (ionization), resulting in an increase of hydrophilicity, and reducing the hydrophilic demulsifier requirement to attain optimum formulation.

Commercial demulsifiers are chemical cocktails or complex mixtures of surfactants (active substance), polymers, polar additives (alcohols, amines, ketones, and others), and solvents. Alcohols are useful substances to promote changes on the interfacial properties of surfactant–oil–water systems [46, 47]. Pei et al. [48] reported correlation between the interfacial tension and solubility in water of alcohols. Interfacial concentration of the surfactant is dependent on the less water-soluble alcohol additives will provide higher alcohol concentrations at the oil –water interface. Generally, the formulation of such products is protected by patents. The mechanisms and modes of action remain undisclosed in open literature. Therefore, the methodology explained constitutes a useful tool for the petroleum industry.

Solvents are used as carriers of the active molecules, and they play a critical role in the performance of a commercial demulsifier. Polar aromatic hydrocarbons are soluble in water, and are frequently used as dissolvent in the formulation of commercial products. Careful consideration is necessary of the solvent is surfactant dispersions then its addition in crude oil is most effective. An inadequate choice of solvent may result in a delay in the rate of breaking of emulsion and therefore other variables of the process physical o chemical are needed in order to improve crude oil desalter operations. Pereira shows that the solvent role can be understood using the criteria of proportional regime with solvents mixtures [7]. The position of optimum concentration of demulsifier (C_D*) is influenced by the nature of the solvent.

5. Conclusions

Here it was explained the approach from the optimum formulation concept to assist emulsion breaking in desalting process. As part of the stabilization mechanism is accepted that the water droplets are stabilized by the formation of a mechanically strong and viscoelastic film at the interface composed of asphaltenes.

In the case of water-in-crude emulsions, a balanced optimum formulation is attained by adding demulsifiers which are hydrophilic to the lipophilic natural surfactants contained in the crude oil. This study aimed to determine an optimal and to relate the nature and concentration of demulsifier products added to the amphiphilic mixture at the interface. Proportional and saturation regimens can explaine the impact of formulation parameters, such as solvent, alcohols, kind and concentration of demulsifier, amoung others as useful tool by the formulating personal of petroleum industry.

Acknowledgements

Juan Pereira is grateful to the projects 2010-03 CDCH-UC and PEII2012-1184 PEI-FONACIT for funding.

Author details

Juan Pereira[1*], Ingrid Velasquez[1], Ronald Blanco[1], Meraldo Sanchez[2], César Pernalete[3] and Carlos Canelón[3]

*Address all correspondence to: jcpereir@uc.edu.ve

1 Laboratory of Petroleum, Hydrocarbon and Derivates, University of Carabobo, Valencia, Venezuela

2 Petróleos de Venezuela S.A. (PDVSA) El Palito Refinery, Carabobo, Venezuela

3 Petróleos de Venezuela S. A. (PDVSA)- Intevep, Gerencia General de Refinación e Industrialización, Gerencia Departamental de Refinación. Los Teques, Caracas, Venezuela

References

[1] Vafajoo L, Ganjian K, Fattahi M. Influence of key parameters on crude oil desalting: An experimental and theoretical study. *J Petrol Sci Engin*. 2012;90-91:107-111. DOI: 10.1016/j.petrol.2012.04.022

[2] Gary J, Handwerck G (eds.). *Petroleum Refining: Technology and Economics*. 4th edn. Marcel Dekker, Inc. New York. USA; 2001. p. 49. ISBN: 0-8247-0482-7

[3] Fahim M, Alsahhaf T, Elkilani A. Crude Distillation. In: Fahim M, Alsahhaf T, Elkilani A, *Fundamentals of Petroleum Refining*. 1st edn. Elsevier B.V. Oxford. UK; 2010. p. 77. DOI: 10.1016/B978-0-444-52785-1.00004-8

[4] Arnold K, Stewart M. Choosing a Process. In: Surface Production Operations. Volume 1. *Design of Oil Handling Systems and Facilities*. 1st edn. Elsevier. B.V. Massachusetts. USA; 2007. p.30. DOI: 10.1016/B978-088415821-9/50011-7

[5] Kang W, Xu B, Wang Y, Li Y, Shan X, An F, Liu J. Stability mechanism of w/o crude oil emulsion stabilized by polymer and surfactant. *Colloids Surfaces A: Physicochem Engin Asp*. 2011;384:555-560. DOI:10.1016/j.colsurfa.2011.05.017

[6] Ramalho J, Lechuga FC, Lucas EF, Effect of the structure of commercial poly(ethylene oxide-bpropylene oxide) demulsifier bases on the demulsification of water-in-crude oil emulsions: elucidation of the demulsification mechanism. *Quím Nova*. 2010;33:1664-1670. DOI: 10.1590/S0100-40422010000800009

[7] Pereira JC, Phenomena of Rupture and Inversion Emulsions: Physicochemical and Kinetic Aspects [Thesis]. Mérida: University of Los Andes; 2009: 11-12, 33, 11, 25-26, 46-47, 64, 46-47, 61-63, 46-47, 77-81.

[8] Silva I, Borges B, Blanco R, Rondón M, Salager J.L, Pereira JC. Breaking of water-in-crude oil emulsions. 5. Effect of acid-alkaline additives on the performance of chemical demulsifiers. *Energy Fuels*. 2014;28:3587-3593. DOI: 10.1021/ef5006918

[9] Kilpatrick P. Water-in-crude oil emulsion stabilization: review and unanswered questions. *Energy Fuels*. 2012;26:4017-4026. DOI: 10.1021/ef3003262

[10] Kralova I, Sjöblom J, Øye G, Simon S, Grimes B, Paso K. Heavy crude oils/particle stabilized emulsions. *Adv Colloid Interf Sci*. 2011;169:106-127. DOI: 10.1016/j.cis.2011.09.001

[11] Daaou M, Bendedouch D. Water pH and surfactant addition effects on the stability of an Algerian crude oil emulsion. *J Saudi Chem Soc*. 2012;16:333-337. DOI: 10.1016/j.jscs.2011.05.015

[12] Fingas M, Fieldhouse B. Studies of the formation process of water-in-oil emulsions. *Marine Poll Bull*. 2003;47:369-396. DOI: 10.1016/S0025-326X(03)00212-1

[13] Fingas M, Fieldhouse B. Studies on water-in-oil products from crude oils and petroleum products. *Marine Poll Bull*. 2012;64:272-283.DOI: 10.1016/j.marpolbul.2011.11.019

[14] Delgado J. Study of Properties of Asphaltene from Boscan Crude Oil [Thesis]. Mérida: University of Los Andes. 2005: 30-32, 12, 30.

[15] Shirani B, Nikazar M, Naseri A, Mousavi-Dehghani S. Modeling of asphaltene precipitation utilizing Association Equation of State. *Fuel*. 2012;93:59-66. DOI: 10.1016/j.fuel.2011.07.007

[16] Batista J, Lechuga F, Lucas E. Effect of the structure of commercial poly(ethylene oxide-b-propylene oxide) demulsifier bases on the demulsification of water-in-crude oil emulsions. *Quim Nova*. 2010;33:1664-1670. DOI: 10.1590/S0100-40422010000800009

[17] Acevedo S, García L.A, Rodríguez P. Changes of diameter distribution with temperature measured for asphaltenes and their fractions A1 and A2. Impact of these measurements in colloidal and solubility issues of asphaltenes. *Energy Fuels*. 2012;26:1814-1819.DOI: 10.1021/ef201947h

[18] Dicharry C, Arla D, Sinquin A, Graciaa A, Bouriat P. Stability of water/crude oil emulsions based on interfacial dilatational rheology. *J Colloid Interf Sci*. 2006; 297:785-791.DOI: 10.1016/j.jcis.2005.10.069

[19] Fingas M, Fieldhouse B. Studies on crude oil and petroleum product emulsions: water resolution and rheology. *Colloids Surfaces A: Physicochem Engin Asp*. 2009;333:67–81.DOI: 10.1016/j.colsurfa.2008.09.029

[20] Pacheco-Sanchez JH, Alvarez-Ramirez F, Martinez-Magadan JM. Morphology of aggregated asphaltene structural models. *Energy Fuels*. 2004;18:1676-1686. DOI: 10.1021/ef049911a

[21] Acevedo S, Castro A, Vasquez E, Marcano F, Ranaudo M.A. Investigation of physical chemistry properties of asphaltenes using solubility parameters of asphaltenes and their fractions A1 and A2. *Energy Fuels*. 2010; 24:5921-5933.DOI: 10.1021/ef1005786

[22] Acevedo S, Ranaudo M.A, García C, Castillo J, Fernández A. Adsorption of asphaltenes at the toluene-silica interface: a kinetic study. *Energy Fuels*. 2003;17:257-261.DOI: 10.1021/ef020104q

[23] Acevedo S, Caetano M, Ranaudo M, Jaimes B. Simulation of asphaltene aggregation and related properties using an equilibrium-based mathematical model. *Energy Fuels*. 2011; 25:3544-3551.DOI: 10.1021/ef200107k

[24] Mullins O, Sheu E, Marshall A (eds.). *Asphaltenes, Heavy Oil and Petroleomic*. Springer. 2007. p. 668. DOI: 10.1007/0-387-68903-6

[25] Acevedo S, Ranaudo MA, Pereira JC, Castillo J, Fernández A, Pérez P, Caetano M. Termo-optical studies of asphaltene solutions: Evidence for solvent-solute aggregate formation. *Fuel*. 1999;78:997-1003. DOI: 10.1016/S0016-2361(99)00027-7

[26] Jeribi M, Almir-Assad B, Langevin D, Hénaut I, Argillier JF. Adsorption kinetics of asphaltenes at liquid interfaces. *J Colloid Interf Sci*. 2002;256:268-272. DOI: 10.1006/jcis.2002.8660

[27] Roodbari H, Badieia A, Soleimani E, Khaniani Y. Tweens demulsification effects on heavy crude oil/water emulsion. *Arabian J Chem*. 2011:1-6. DOI: 10.1016/j.arabjc.2011.08.009

[28] Nenningsland A, Gao B, Simon S, Sjöblom J. Comparative study of stabilizing agents for water-in-oil emulsions. *Energy Fuels*. 2011;25:5746-5754. DOI: 10.1021/ef2014265

[29] Gao S, Moran K, Xu Z, Masliyah J. Role of naphthenic acids in stabilizing water-in-diluted model oil emulsions. *J Physic Chem B*. 2010; 114: 7710-7718. DOI: 10.1021/jp910855q

[30] Sullivan P, Zaki N, Sjöblom J, Kilpatrick P. The stability of water-in-crude and model oil emulsions. *Canadian J Chem Engin*. 2007; 85: 793-807. DOI: 10.1002/cjce.5450850601

[31] Yarranton H, Hussein H, Masliyah JH. Water in-hydrocarbon emulsions stabilized by asphaltenes at low concentrations. *J Colloid Interf Sci*. 2000; 228:52-63. DOI: 10.1006/jcis.2000.6938

[32] Gafonova O, Yarranton H. The stabilization of water-in-hydrocarbon emulsions by asphaltenes and resins. *J Colloid Interf Sci*. 2001;241:469-478. DOI: 10.1006/jcis.2001.7731

[33] Silva I. Breaking of Water-in-Crude Oil Emulsions: Effect of Acid-Alkaline Additives on the Performance of No Ionic Demulsifiers. [Thesis]. Caracas: University Simón Bolívar. 2012:24.

[34] Zhang L, Breen P, Xu Z, Masliyah JH. Asphaltene films at a toluene/water interface. *Energy Fuels*. 2007;21:274-285. DOI: 10.1021/ef0603129

[35] Delgado-Linares J, Pereira JC, Rangel A, Salazar F, Guevara M, Forgiarini A, Bullón J. Effect of addition of amphiphilic molecules on asphaltene film properties at air-water interface. *Revista de la Facultad de Ingeniería de la UC*. 2012;19:16-24. ISSN 1316-6832

[36] Solovyev A, Zhang L, Xu Z, Masliyah LY. Langmuir films of bitumen at oil/water interfaces. *Energy Fuels*. 2006;20:1572-1578. DOI: 10.1021/ef050409f

[37] Chandra M, Xu Z, Masliyah JH. Interfacial films adsorbed from bitumen in toluene solution at a toluene–water interface: a Langmuir and Langmuir–Blodgett film approach. *Energy Fuels*. 2008;22:1784-1791. DOI: 10.1021/ef700786u

[38] Zhang LY, Lawrence S, Xu Z, Masliyah JH. Studies of Athabasca asphaltene Langmuir films at air–water interface. *J Colloid Interf Sci*. 2003; 264:128-140. DOI: 10.1016/S0021-9797(03)00355-2

[39] Zhang L.Y, Xu Z, Masliyah JH. Characterization of adsorbed Athabasca asphaltene films at solvent-water interfaces using a Langmuir interfacial trough. *Indust Engin Chem Res*. 2005;44:1160-1174. DOI: 10.1021/ie049913i

[40] Rosen MJ (ed.). *Surfactants and Interfacial Phenomena*. 3rd edn. Wiley-Interscience, 2004; p. 455. ISBN 0-471-47818-0

[41] Balsamo V, Phan J, Nguyen D. Interfacial films of diluted heavy oil–aqueous systems at elevated temperatures and pressures: effect of reverse emulsion breakers. *Energy Fuels*. 2015;29:2823-2832. DOI: 10.1021/ef502072f

[42] Rondón M, Bouriat P, Lachaise J, Salager JL. Breaking of water-in-crude oil emulsions. 1. Physicochemical phenomenology of demulsifier action. *Energy Fuels.* 2006;20:1600-1604. DOI: 10.1021/ef060017o

[43] Rondón M, Pereira JC, Bouriat P, Graciaa A, Lachaise J, Salager JL. Breaking of water-in-crude-oil emulsions. 2. Influence of asphaltene concentration and diluent nature on demulsifier action. *Energy Fuels.* 2008;22:702-707. DOI: 10.1021/ef7003877

[44] Pereira JC, Delgado-Linares J, Scorzza C, Rondón M, Rodríguez S, Salager JL. Breaking of water-in-crude oil emulsions. 4. Estimation of the demulsifier surfactant performance to destabilize the asphaltenes effect. *Energy Fuels.* 2011;25:1045-1050. DOI: 10.1021/ef100979y

[45] Pensini E, Harbottle D, Yang F, Tchoukov P, Li Z, Kailey I, Behles J, Masliyah JH, Xu Z. Demulsification mechanism of asphaltene-stabilized water-in-oil emulsions by a polymeric ethylene oxide–propylene oxide demulsifier. *Energy Fuels.* 2014;28:6760-6771. DOI : 10.1021/ef501387k

[46] Baviere M, Schechter R, Wade W. The influence of alcohols on microemulsion composition. *J Colloid Interface Sci.* 1981;81:266-279. DOI: 10.1016/0021-9797(81)90322-2

[47] Pereira JC. State of the art on the effect of alcohols on the interfacial properties of surfactants. *Revista Ingeniería* UC. 2012; 19:76-85. ISSN 1316-6832

[48] Pei H, Zhang G., Ge J, Zhang L, Mingchao MA. Effect of the addition of low molecular weight alcohols on heavy oil recovery during alkaline flooding. Ind Eng Chem Res. 2014; 53:1301-1307. DOI: 10.1021/ie4028318

Chapter 5

Overview about Different Approaches of Chemical Treatment of NORM and TE-NORM Produced from Oil Exploitation

N.S. Awwad, M.F. Attallah, E.M. El-Afifi,
H.A. Ibrahium and H.F. Aly

Additional information is available at the end of the chapter

http://dx.doi.org/10.5772/61122

Abstract

Crude oil or gas is considered the most important sources of natural radionuclides from the uranium and thorium series. The radionuclides of primary concern to the oil and gas industry are Ra226 (U238 decay) and Ra228 (Th232 decay) due to their radiotoxicity and relatively long half-lives (1620 and 5.75 years, respectively). One of the decay products of U238 is Rn222, a radioactive noble gas. Radon emanates from the ground where it can be breathed in. Radon also decays to alpha-emitting progeny nuclides, so inhaling a single radon atom can lead, ultimately, to up to four alpha decays in the lungs before reaching stability. Alpha particles are considered as internally hazardous if a radioactive source emitting its particles is inhaled or ingested. So, in this chapter we shall make an overview about different approaches of chemical treatment of NORM and TE-NORM produced from oil exploitation.

Keywords: Chemical Treatment / NORM , TE-NORM/ Oil Exploitation / Scales / Sludge

1. Introduction

Most of the natural radionuclides in wastes from crude oil exploitation were firstly discovered in the beginning of the last century. More types of industries have been identified dealing with

INTECH
open science | open minds

© 2015 The Author(s). Licensee InTech. This chapter is distributed under the terms of the Creative Commons Attribution License (http://creativecommons.org/licenses/by/3.0), which permits unrestricted use, distribution, and reproduction in any medium, provided the original work is properly cited.

materials containing enhanced levels of natural radionuclides, which are summarized as TE-NORM (technologically enhanced naturally occurring radioactive materials). Naturally occurring radionuclides are present at varying concentrations in the Earth's crust and can be concentrated and enhanced by processes associated with the recovery of oil and gas. This "enhanced" NORM, often known as TE-NORM, can be created as a result of industrial processes and human activity, and in addition, can be the by-product of oil, gas production. One of the important examples that can contain elevated levels of NORM, and the radioactive materials may migrate from site to site as the materials and equipment are reused are sludge, pipe scales, produced water, and drilling mud.

Transportation or decay of radioactive elements produces other radionuclides (daughters) from the reservoir to the surface with the produced oil and gas carried out under certain conditions dependent upon pressure, temperature, acidity. NORM with the oil, gas, and water mixture migrate and accumulate in scale, sludge, and scrapings during the production process. Moreover, they form a thin film on the interior surfaces of gas-processing vessels and equipment. The geological formation is considered as important parameters at determining the level of NORM and it was found that the accumulation can vary from one site to another depending on it.

There are three types of radiation emitted by NORM, namely:

· Alpha (α) · Beta (β) · Gamma (γ)

Alpha particles are considered as internally hazardous if a radioactive source of alpha-emitting particles is inhaled or ingested. While beta particles have one (negative) charge and interact more slowly with the material, they are effectively stopped by thin layers of metal, wood, or plastic and considered hazardous only if a beta-emitter source is ingested or inhaled. Gamma emitters are associated with alpha, beta decay with high-energy electromagnetic radiation that interacts lightly with matter. Gamma rays are best shielded by thick layers of lead or other dense materials and are considered as an external hazard to human bodies. Figure 1 details the origins of NORM in the recovery process [1].

2. Natural Radionuclides in the Oil Industry

2.1. Radon

One of the decay products of U238 is Rn222, a radioactive noble gas. Radon emanates from the ground where it can be breathed in. Radon also decays to alpha-emitting progeny nuclides, so inhaling a single radon atom can lead, ultimately, to up to four alpha decays in the lungs before reaching stability. Since radon comes from the decay of geologic uranium, radon concentrations will vary according to local geology. In general, radon dose will be higher in areas that are underground, poorly ventilated, and in areas with high levels of uranium in the rocks.

Crude oil or gas is considered the most important sources of natural radionuclides from the uranium and thorium series. The radionuclides of primary concern to the oil and gas industry

http://dx.doi.org/10.5772/61122

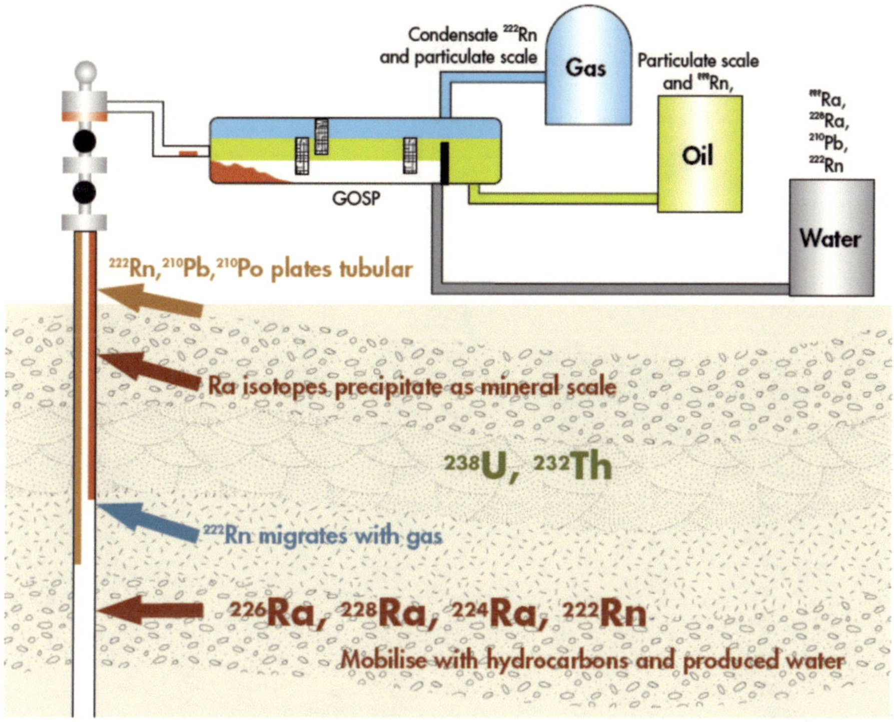

Figure 1. The origins of NORM in the recovery process [1]

are Ra226 (U238 decay) and Ra228 (Th232 decay) due to their radiotoxicity and relatively long half-lives (1620 and 5.75 years, respectively) as shown in Figures 2 and 3 [1]. Radon radionuclides escaping from the adjacent geological formations are soluble in crude oil, but due to its half-life (3.825 days) only Rn222 is

present in the pumped oil in varying concentrations from 10 to 800 Bq/kg [1]. Many previous data on the distribution of the main radon nuclide Rn222 in different gas fields and processing plants were collected in the United Nations Scientific Committee on the Effects of Atomic Radiation (UNSCEAR) report [2]. Crude oil is usually pumped to the surface together with produce water with radon and radium radionuclides while uranium and thorium usually do not go into solution. Various cations such as barium and strontium, with anions such as sulfate, chloride, or the bicarbonate solubility of radon in water is lower, and the level of Rn222 concentrations in the petroleum formation water was found in the range of 18.5 Bq/dm^3 [3]. Very low concentrations measured for the U and Th and produced waters were obtained under the reducing conditions. The same result was found with radium nuclides released by alpha from the surrounding minerals or leaching processes usually recovered by sorption, except for saline waters with high concentration of chloride anions [4–5]. Typical ranges or average values of the radium radionuclide concentrations in the formation or produced water from different oil fields, including the recent data, are listed in Table 1.

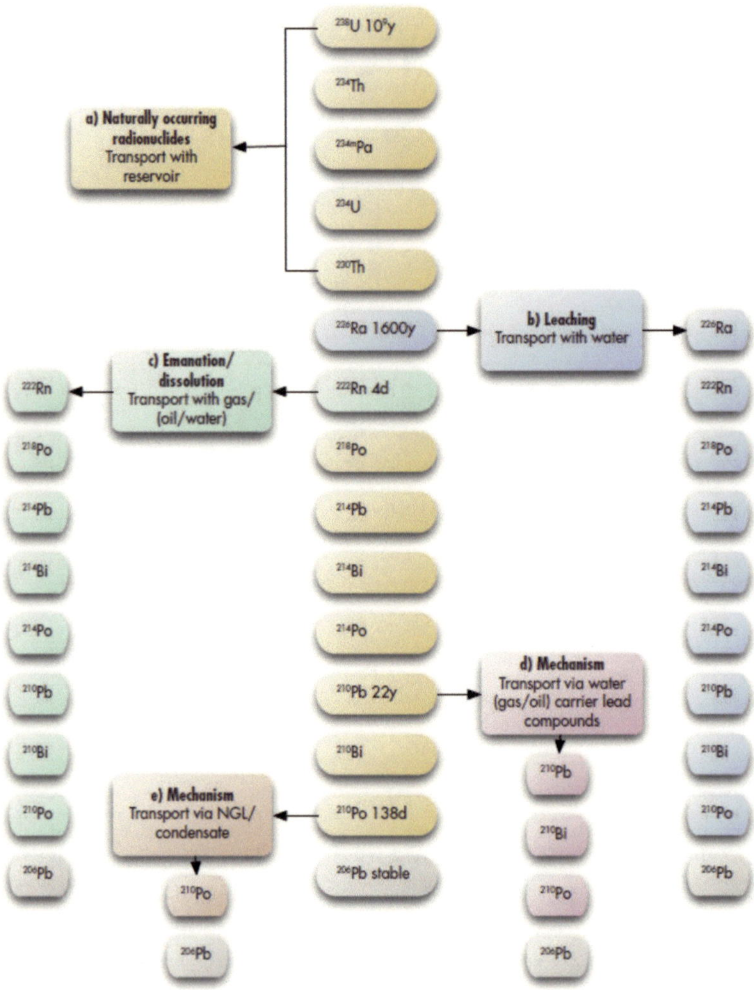

Figure 2. Uranium238 decay series [1]

The worldwide average concentration of these radionuclides in produced water discharged to the environment is estimated at 10 Bq/l. These concentrations are approximately three orders of magnitude higher than the natural concentrations of radium in drinking or sea water.

Scale formation is a complex phenomenon and can be interpreted by the variation of the solubility of carbonates or sulfates or by pressure and temperature changes, water injection into the reservoirs, and evaporation in the gas extraction pipes. Radium and radon concentrations in the pipe scale and waste sludge are dependent on three factors: the amount of Ra present in the subsurface soil, treatment processes applied during oil or gas production, and formation of water components. Most radium radionuclides are efficiently concentrated from the water phase during formation of the scale. Therefore, the measured levels of activity concentrations both in the separated sludge and in the solid scale are much higher than those

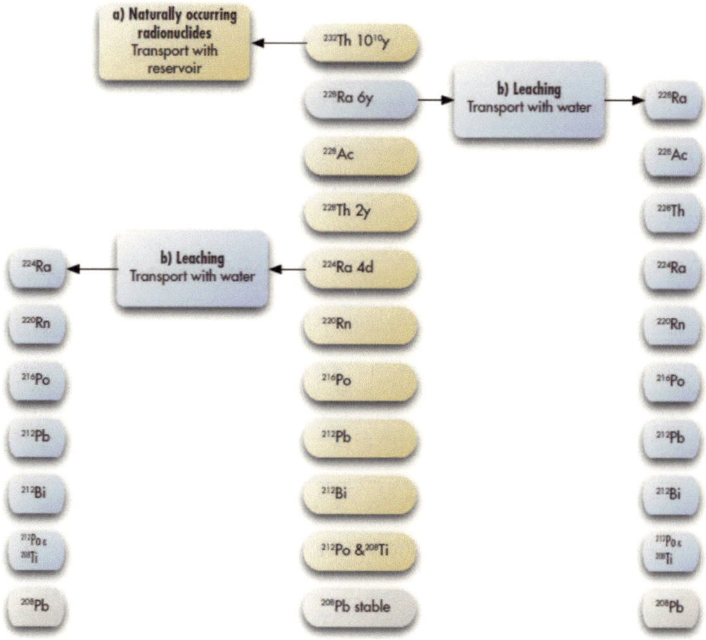

Figure 3. Thorium232 decay series [1]

measured in the produced water from the oil industry. Ra226 and Ra228 activity concentrations measured in the solid scale and sludge are listed in Table 2.

^{228}Ra (Bq/dm^3)	^{226}Ra (Bq/dm^3)	Sample	Field
	1000–950.000	Scale	Algeria [8]
	21.000–250.000	Scale	Australia [9]
	19.100–323.000	Scale	Brazil [20]
48.000–300.000	121.000–3.500.000	Scale	Brazil [21]
4210–235.000	77.900–2.110.000	Scale	Brazil [22]
148.000–2.195.000	97–151	Scale	Congo [11]
101.500–1.550.000	68.900	Scale	Egypt [23]
24.000	7541–143.262	Scale	Egypt [12]
35.460–368.654	< 2.7–2890	Scale	Italy [11]
200–10.000	510–51.000	Scale	Kazakhstan [24]
130.120–206.630	114.300–187.750	Scale	Malaysia [25]
300–33.500	300–32.300	Scale	Norway [26]
30.000	.08–1.5	Scale	Saudi Arabia [27]
	31–1189	Scale	Tunisia [11]
	4300–658.000	Scale	Tunisia [28]
	1000–1.000.000	Scale	UK [29]
	up to 3.700.000	Scale	USA [30]

^{228}Ra (Bq/dm^3)	^{226}Ra (Bq/dm^3)	Sample	Field
	5.1–14.8	Formation water	Algeria [7–8]
23[a]	17[a]	Produced water	Australia [9]
0.05–12	0.01–6	Produced water	Brazil [10]
1–59	5.1[a]	Produced water	Congo [11]
-	5–40	Formation water	Egypt [12]
2.8[a]	0.2-2	Produced water	Italy [11]
0.5–21	0.3–10.4	Formation water	Norway [12]
8.8–60.4	3.3[a]	Produced water	Norway [13]
-	0.5–16	Produced water	Norway [14]
-	9.9-111.2	Produced water	Syria [15]
0.7–1.7	1.7[a]	Produced water	UK [16]
15.1[a]	0.1–60	Produced water	USA [6]
25–30	0.15–21.6	Produced water	USA [17]
	12.6[a]	Oilfield brine	USA [18]
	22–30	Produced water	USA [19]

a: Mean activity concentration.

Table 1. Ranges of activity levels in produced water from the oil fields

^{228}Ra (Bq/dm^3)	^{226}Ra (Bq/dm^3)	Sample	Field
	15.400–76.100	Scale	USA [31]
	25.000	Sludge	Australia [9]

Table 2. Ranges of activity levels of ^{226}Ra in different scale and sludge samples

From the data obtained in Table 2, it was found that the concentration levels of radium nuclides in sludge were much lower than those of the scale. Environmental Protection Agency (EPA) estimated the average radium nuclide concentration to be around 2800 Bq/kg and 18, 000 Bq/kg in sludge and scale, respectively [32]. The exposure is caused by external radiation coming from the 226Ra radionuclide and its progenies: Pb214 and Bi214 as well as by inhalation of α-emitting radionuclides; Rn222 as well as Po218 and Po214 formed from Rn222 escaping into the air adjacent to scale deposits, see Table 3.

The main types of scales encountered in oil and gas facilities are sulfate scale such as $BaSO_4$, which is called barite. Colorless or milky white is the common one, but can have any color, depending on the impurities trapped in the crystals during barite formation. The high density of barite is responsible for its value in many applications. It is always chemically inert and

Reported range (μSv/h)	Country
Background–100	Algeria [8]
10–300	United Kingdom [8]
50–100	Egypt [23]
0.1–6	Congo, Italy, Tunisia [11]
up to 300	USA [5]

Table 3. Exposure rate levels in the oil industry

insoluble. Due to the high chemical similarity of radium with barium (Ba), strontium (Sr), and calcium (Ca), radium co-precipitates with Sr, Ba, or Ca scale forming radium sulfate, radium carbonate, and in some cases radium silicate is produced. As mentioned before, the mixing of seawater, which is rich in sulfate, with the produced water, which is rich in Cl^{-1}, increases the scaling tendency. In addition, any change in pressure and temperature or acidity of the formation water contributes to scale build-up. The build-up of scale on the interior of a pipe is shown in Figure 4. The activity concentrations of Ra226 and Ra228 in hard scales in Egypt and some other countries were mentioned in Table 4.

Figure 4. The build-up of scale on the interior of a pipe

Inhibitors may be applied to the piping complexes to prevent scales from slowing the oil extraction process. If the scales contain TE-NORM, the radiation will remain in solution and eventually be passed on to the produced waters. The United States generated annually around 100 tons of scale per oil well. Sometimes, in some cases brine is introduced into the produced water to enhance the recovery of radium; this also increases scale formation.

Country/region	Activity (Bq/g)		Ref.
	^{226}Ra	^{228}Ra	
Egypt/oil field			
Abu Rudeis	68.9	24	[33]
Gabal El Zeit	14.8	4.3	[34]
Badr El Din	31.4	43.3	[34]
Red Sea	195	897.8	[35]
Western desert	59.2	244.5	[35]
Other countries			
Australia	20–70		[36]
USA	70.8		[37]
Algeria	1–950		[38]
Tunisia	4.3–658		[39]
UK	1–1000		[40]

Table 4. Activity of ^{226}Ra (U-series), ^{228}Ra (U-series), and ^{40}K in the TE-NORM in Egypt and some countries

2.2. NORM in sludge

Radioactive molecules containing radium which were not incorporated into scale can be found in sludge, produced sands, and produced waters. Before the treatment investigations, the activity concentrations of the main three radium isotopes were measured in sludge [41]. It is found that the average activity level of Ra226, Ra228, and Ra224 for samples taken from Abu Rudeis region, Government of Sinai–Egypt, were 11950, 1750, and 1900 Bq/kg, respectively. Due to the accumulated TE-NORM wastes in huge amounts and release of high activity concentrations to the environment and the workers, treatment approaches are suggested to reduce the human and environmental hazard potential. Sludge is composed of dissolved solids which precipitate from produced water as its temperature and pressure change. Sludge generally consists of oily, loose material often containing silica compounds, but may also contain large amounts of barium.

2.3. NORM in natural gas

Radon is a radioactive noble gas, present in varying level in natural gas in oil and gas formations. In the absence of natural gas, radon easily dissolves in the light hydrocarbon. Radon usually follows the gas stream when produced with the oil and gas. If the natural gas is fractionated, higher percentage of radon can concentrate in the propane streams and lower degree in the ethane streams. Most of the radon decay products are found in ambient aerosols, airborne particulates, or surfaces. Most radon progeny are short-lived, with the exception of Po210 and Pb210, which have relatively long half-lives of 138 days and 22.6 years, respectively. Formation of thin radioactive films on the inner surfaces of gas-processing equipments, such as scrubbers, compressors, reflux pumps, control valves, and product lines, due to the release of daughters of radon.

2.4. NORM in produced water

Formation water at oil and gas reservoirs called produced water float to the surface during oil and gas production. However, gas reservoirs typically produce smaller quantities whereas the oil reservoirs can contain large volumes of formation water. Water is injected into the reservoir to maintain pressure and or maximize production in most fields. The produced water stream can be one of the largest waste products, by volume produced from oil and gas industry. It was found that the produced water contains a complex mixture of inorganic compounds, such as dissolved salts, trace metals, suspended particles and organic compounds such as dispersed and dissolved hydrocarbons, organic acids, and residual chemical additives such as scale and corrosion inhibitors that are added into the hydrocarbon production process. Technologies of treatment depending on the application and particular field conditions. Also, it is to consider include combinations of gravity and / or mechanical separation and chemical treatment, and may include a multistage system, typically including a parallel plate separator, followed by a gas flotation cell or hydrocyclone. For disposal of produced water to sea, some criteria should be considered to reduce the volume of produced water, such as:

Select adequate well management during well completion to minimize water production;

Minimize water production by recompletion of high water-producing wells;

To be economically feasible, use down hole fluid separation techniques, where possible.

For reducing environmental hazards related to residual chemical additives in the produced water stream, where surface disposal methods are used, production chemicals should be selected carefully by taking into account their volume, toxicity, and bioaccumulation potential [42]. The average worldwide activity levels of uranium (U), thorium (Th), and potassium (K) [43] and the exemption activity levels of NORM as recommended in the IAEA basic safety standards [44] were given in Tables 5 and 6. The average worldwide levels of the most common radiological indices [43] were given in Table 7. These indices include radium equivalent (Ra-eq), total absorbed dose ($D_{\gamma r}$), and effective annual dose rate (EDAR).

Radionuclide	U	Th	K
Activity level (Bq/Kg)	50	50	500

Table 5. The average worldwide activity levels of uranium (U), thorium (Th), and potassium (K)

Radionuclide	U^{238}	Ra^{226}	Rn^{222}	Th^{232}	Ra^{228}	Ra^{224}
Exemption level (Bq/g)	1	10	10	1	10	10

Table 6. The exemption activity levels of NORM as recommended in the IAEA basic safety standards

Radiological indices (Unit)	Ra-eq (Bq/Kg)	Dγr (nGy/h)	EADR (mSv/yr) for worker	EADR (mSv/yr) for Public
Activity level (Bq/Kg)	370	55	20	1

Table 7. The average worldwide levels of the most common radiological indices

3. Resources of NORM and TE-NORM

Earlier, we saw that many natural materials contain radioactivity. Some of these materials are used in manufacturing: clays and various minerals. In other cases, radioactive elements are used in manufacturing – not because of the radioactivity, but because of other chemical or physical properties. For example, thorium is radioactive, and it has a high melting temperature; for this reason, thorium is used in the manufacture of some welding electrodes, gas lantern mantles, and jet turbine blades. These products and the wastes from their manufacture will contain low levels of radioactivity. Many fossil fuels are associated with radioactivity; the geochemistry of uranium is such that it is often found in petroleum, natural gas, and coal deposits. The equipment used to extract and process these materials is often contaminated with NORM materials, especially with radium, which can be present in the scales and sludges from processing these materials. Although refined petroleum and natural gas products do not contain large levels of these nuclides, coal can; so, fly ash often contains elevated levels of radioactivity as well. In fact, when 90% or more of the coal is burned, all of the radioactivity is left in the remaining 10% that is ash. Thus, fly ash is considerably more radioactive than is the original coal. Uranium and thorium are fairly common in nature, and many minerals contain elevated levels of these radioactive elements and their decay series nuclides. In particular, minerals containing rare earth elements (such as monazite), titanium minerals, niobium ores, and some precious metal ores can be associated with elevated levels of radioactivity. When the minerals are processed, the metal is removed from the ore, and the radioactivity concentration in the remaining waste is even higher than in the original ore. In addition to all of these, anything that contains potassium will be somewhat radioactive because 1/100% of potassium is naturally radioactive. Potassium is found in many minerals, some forms of clay, and in many products – consumer and industrial both – so sometimes these items will also give high levels of radiation. Loads of bananas, kitty litter, and salt substitute all contain enough radioactive potassium to set off radiation alarms.

3.1. NORM and TE-NORM at the landfill

Many industries produce wastes that might contain natural radionuclides, and most of these wastes end up at landfills, either hazardous or otherwise. The majority of TE-NORM sources waste arise from industrial processes. Most of the wastes obtained are produced in very large amounts with low activity. Improper disposal, recycling, and reuse of TE-NORM have led to circumstances resulting in contamination and unnecessary public exposures.

Disposal of TE-NORM in piles or stacks can lead to contamination of groundwater and airborne releases of radioactive particulates and radon. Improper use and/or disposal of the TE-NORM used for soil conditioning or fill around homes can lead to accumulation of radon gas in homes, direct exposure to individuals, and contamination of soil and of the crops growing in the soil. Reuse of TE-NORM-contaminated materials, such as in concrete aggregate, ceramic industry, and red stone can lead to increased radiation risks to members of the public in a variety of ways.

The main radionuclides daughter found in the uranium series in industrial TE-NORM are ^{238}U, ^{234}U, ^{230}Th, ^{226}Ra, and ^{222}Rn, while the thorium series are ^{232}Th, ^{228}Ra, and ^{220}Rn. In addition, ^{40}K should be characterized. Radium226 is used as a monitor to show the relative activity and volume among the TE-NORM sectors. It was found that Radium226 is in equilibrium with its progeny, but Radium228 is not and their activity appears to be about three times greater than Radium228.

4. General Technologies for Treatment of TE-NORM Waste

4.1. Use of reagents

A water-bath heater from a gas production facility in North Sea (e.g., the Netherlands) has been used as a test for equipment in laboratory aimed at the in situ removal of TE-NORM. This is achieved by circulating an aqueous solution of commercially available scale dissolver through the contaminated equipment. For this purpose, some scale dissolvers which are widely used within exploration and production are commonly based on chelating chemistry and reportedly successfully applied in the dissolution of low specific activity scales [45]. For instance, application of scale dissolver reagents resulted in the rapid and complete removal of ^{226}Ra and progeny-containing sulfate scales as well as ^{210}Pb-containing sulfide scales from the head internals. These studies were performed using scale dissolver consisting of 15% v/v acetic acid and 1% v/v strongly oxidant, e.g., $KMnO_4$. This reduced the total activity from 20 Bq/cm^2 to 6 Bq/cm^2, the residual activity may be removed using 0.5 M citric acid. Also, scale dissolver solution containing 15% v/v acetic acid plus 1% v/v hydrogen peroxide was used to dissolve TE-NORM contaminated by overall activities of 2000 Bq (^{226}Ra)/g and 600 Bq(^{2t0}Pb)/g. This was followed by water flush. Generally, the application of a chemical scale dissolver can remove 95–99% of TE-NORM present in exploration and production facilities. On the other hand, radioactive scales containing ^{226}Ra and its progeny, such as barium sulfate scales, are removed chemically using hydrochloric acid, and the dead acid is disposed as waste after appropriate radiometric checks [46].

4.2. Recycling equipment

The contamination produced from the accumulation of toxic materials mainly include heavy metals such as mercury and radioactive materials of the natural origin (e.g., ^{226}Ra, ^{228}Ra, ^{210}Pb). For decontamination purpose, the contaminated equipments were taken into melting plant, especially built for this process; its annual capacity is 2000 tones of steel and metal scrap contaminated with mercury and TE-NORM. After melting, the radiological measurements showed that the produced metal did not contain any detectable residual of TE-NORM, and can be re-used again in steel works. About 98% of TE-NORM were bound to the slag and ~2% were detected in the filter dust, mainly consisting of the nuclides ^{210}Pb and ^{210}Po. The secondary waste produced is ~43% of the total weight of the material supplied, whereas TE-NORM waste consists of ~95% of slag and ~5% coarse dust [45].

Chemical separation of the radionuclides incorporated in the contaminated equipment (pipelines, tubes, pumps) is carried out by melting at 1400°C, to further fractionation of radionuclides in melt, slag, or dust. The analysis of data showed that most of ^{238}U and ^{232}Th series are transferred from melt (dense main component, contains only 1% of the remainder radioactivity) into the slag (light minor component, contains only 98% of the total radioactivity). All activity of ^{210}Pb was concentrated in the filter dust, because it is evaporated at normal melting temperature above 1300°C [47]. Equipment should be decontaminated to less than 0.4 Bq/cm^2 for alpha emitters or 4 Bq/cm^2 for beta and gamma emitter, before any release.

4.3. Solid TE-NORM waste

For TE-NORM-contaminated scale, sludge, and soils with very low levels of radioactivity, a suitable disposal option is to spread over the ground and mix with non contaminated soils, to dilute the contaminated soils and reduce the radioactivity level to background levels. This type of disposal is often the most cost-effective [48]. Subsurface disposal options include underground injection and down hole encapsulation. This type of disposal is widely acknowledged as one of the most environmentally sound methods of disposing TE-NORM-contaminated sludge. The two common forms of subsurface disposal are:

i. Underground injection, established by mixing a TE-NORM-contaminated waste with cement in a slurry, then injecting the formed mixture into a deep subsurface formation.

ii. Down hole encapsulation, entails placing TE-NORM-contaminated scale, sludge, tubes, and other small pieces of the production equipment (e.g., valves, filters, pumps, screens) inside the casing of a well, which is to be plugged with cement and then abandoned.

4.4. Liquid TE-NORM waste

TE-NORM in slurry form (e.g., waste water or solids mixed with water) can be re-injected into deep formations for disposal [18]. There are three classes for injection:

Class (I): This option is used for any liquid TE-NORM wastes. Over 90% of all produced water resulting from oil and gas operations is injected through wells into permeable disposal formations, which lie below underground sources of drinking water (USDW), and surrounded by impermeable layers. After injection, the well is closed, sealed with cement, and capped, effectively isolating injected materials from the surface. Injection costs vary based on volume, depth, formation pressure and permeability, and other factors. The cost of injecting a slurry could be comparable, or slightly higher.

http://dx.doi.org/10.5772/61122

Class (II): Well injection, when TE-NORM concentrations prevent disposal in class (I). The used wells in this class are deeper and are constructed to give great protection against potential migration of injected fluids to (USDW). Disposal in class (II) well is to some degree more expensive than class (I) injection. Also, transportation costs would be higher, as limited number of class (II) disposal wells exist.

Class (III): Deep well injection, these wells consist of injecting liquid wastes contaminated by TE-NORM fluids into the well at sufficiently high pressure to create a fracture in permeable shale formations. After the scale/water mixture is displaced into the fracture, then the pressure is reduced, and the fracture closes.

The scale is trapped between the fracture wells and is incapable of re-entering the well bore. Deep well injection is generally regarded as an effective method for the disposal of TE-NORM waste because it does not depend on the mechanical integrity of the well to prevent potential subsurface contamination.

When the radium ions are present in the produced water, any drops in pressure and temperature can lead to the solubility products of their sulfates and carbonates being exceeded. This is the main cause for precipitation of radium as sulfate and carbonate scales on the inner walls of production tubules, well heads, valves, pumps, separators, water treatment vessels, gas treatment and oil storage tanks. Particles of clay or sand co-produced from the reservoir may also act as catalytic surfaces for initiating scale deposition or may adsorb the cations. Daifullah and Awwad [49] found that oil shale is a good adsorbent for Hg(II) from aqueous solution. Shales normally contain at least 35% clay minerals, and a significant fraction contains potassium as an essential constituent. Removal of mercury (II) from wastewater was studied using camel bone charcoal [50]. Shales can adsorb the series radionuclides [51]. Common anthropogenic sources of mercury include nuclear fuel production as part of the uranium purification and isotope separation process (^{235}U and ^{238}U). Mercury in the form of Hg(II) also enters aquatic environments from industrial and nuclear fuel wastes. The feasibility of using oil shale for removal of Hg(II) has been addressed. Also, it was found when using seawater to enhance oil recovery, it will increase the sulfate concentration of the produced water and enhance scale deposition. So, new trends should be used to solve these problems.

5. New Approaches Used in the Treatment of NORM and TE-NORM Produced from Sludge Waste Obtained from Oil and Gas Production

In oil and gas production, the TE-NORM may be solid waste as scale and/or sludge and produced water. Our literature shows that TE-NORM contains activity concentrations of Ra226 ranging from undetectable levels to 1000 kBq/kg [52] and the activity level concentrations of Ra226 in TE-NORM can be much higher than the exemption levels established by IAEA [53]. As mentioned before, the recommended exemption level for uranium-series is: U238 = 1 Bq/g and Ra226 = 10 Bq/g. For thorium decay, a chain is: Th232 = 1 Bq/g, Ra228 = 10 Bq/g, and Ra224= 10 Bq/g.

It is important to focus on the environmental and health impacts from the uncontrolled release of TE-NORM wastes [54, 34, 23]. Treatment of these wastes is of increasing interest because accumulation of large amounts with a significant activity may cause health risks to the workers through exposure, inhalation of radon (Rn222) decayed from radium, and/or ingestion of waste dust during the periodical maintenance of the equipment used. Treatment of TE-NORM wastes from many industries still needs more efforts. The traditional methods used before include subsurface disposal, volume reduction, use of scale and/or sludge inhibitors, recycling, and leaching using chemical solutions [55–57]. In addition, a simple extraction process is carried out using saline solutions and chemical solutions [41, 58] to removal of Ra226, Pb210, Rn220, Th232, Ra228, and Ra224 from TE-NORM wastes produced from oil and gas industry.

Sequential chemical treatment for radium in sludge or scale to reduce its activity concentration in oil and natural gas production fields is recommended. The proposed treatment method was carried out on the basis of two approaches using chemical solutions through four successive steps.

Successive four steps were used to leachate the radium species in the waste of TE-NORM [59].

Before the treatment investigations, the activity concentrations of the main three radium isotopes were measured. It is found that the average activity level of Ra226, Ra228, and Ra224 were 11950 ± 1700, 1750 ± 200, and 1900 ± 250 Bq/kg, respectively. Due to the high accumulation of radium species in huge amounts and high activity concentrations causing health hazards to the environment and the workers, sequential chemicals treatment approaches are suggested as a new trend to reduce the human and environmental hazard potential.

5.1. Approach of treatment

It is well known that the environmental behavior and toxicity of trace elements and radionuclides depend strongly on their physicochemical forms (i.e., speciation) in the environment [60]. In this study, the applied treatments involve four steps achieved sequentially for each approach. Selective extraction of the different radium species present in TE-NORM waste, such as water-soluble species, exchangeable, carbonates, reducible species, oxidizable organics becomes allowed.

5.1.1. Approach 1

The treatment of the radium content in TE-NORM using sequential chemical leaching was based on the individual extraction for each Ra species in the waste, according to the successive four steps (A.1–4). From the data obtained, it was found that the exchangeable radium species was removed from the waste. The removal percentages (%) for Ra226, Ra228, and Ra224 are 5.7 ± 2.4, 6.5 ± 1.4, and 3.1± 0.9%, respectively. These values are high if comparable to the exchangeable Ra species present in and extracted from phosphate ores [61].

In the second step of leaching (A.2), the data obtained show that the removal percentages (%) are found to be of 9.9 ± 0.4, 7.5 ± 0.9, and 11.8 ± 0.2 % for Ra226, Ra228, and Ra224, respectively. This leached part is related to the Ra fraction bounded to carbonate species (acidic fraction of species Ra). In step number three (A.3), it was found that the removal percentages of Ra226, Ra228, and Ra224 are 10.9 ± 1.4, 18.3 ± 2.5, and 19.6 ± 0.4%, respectively. This leached part is related to radium species bounded to metal-oxides such as the Fe-Mn-oxides [62]. The remaining part of radium species found bounded to organic matter and sulfides was leached through two substeps (A.4). The final removal percentages by this approach related to Ra226, Ra228, and Ra224 are 51.5 ± 2.1, 32.5 ± 4.1, and 41.9 ± 5.2%, respectively, as shown in Figure 5.

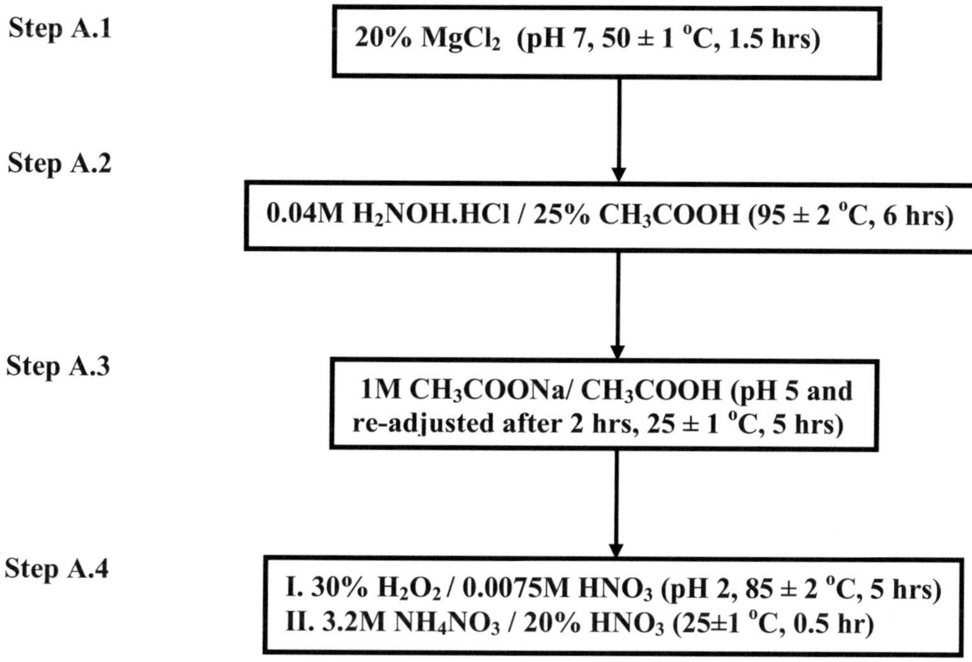

Figure 5. Sequential leaching of the radium content in TE-NORM was based on the individual extraction for each Ra species in the waste

According to approach 1, the successive leaching steps released most of the radium species found in the treated TE-NORM waste. Also, from the data obtained, it is observed that the real removal percentages (%) of Ra226, Ra228, and Ra224 are 78 ± 2.8, 64.8 ± 4.1, and 76.4 ± 5.2%, respectively. There is variation in the leaching % for each Ra-isotope due to the radiochemical factors such as the differences in their half-lives. Figure 6 shows the leaching of the different Ra species in the waste. It is observed that the oxidizable Ra species is the main Ra fraction in this type of waste. This may be attributed to the high concentration of leaching solutions used to remove Ra species within the TE-NORM sludge waste. Therefore, the radium species in the treated waste using approach A can be ordered as: oxidizable > reducible > acidic > exchangeable.

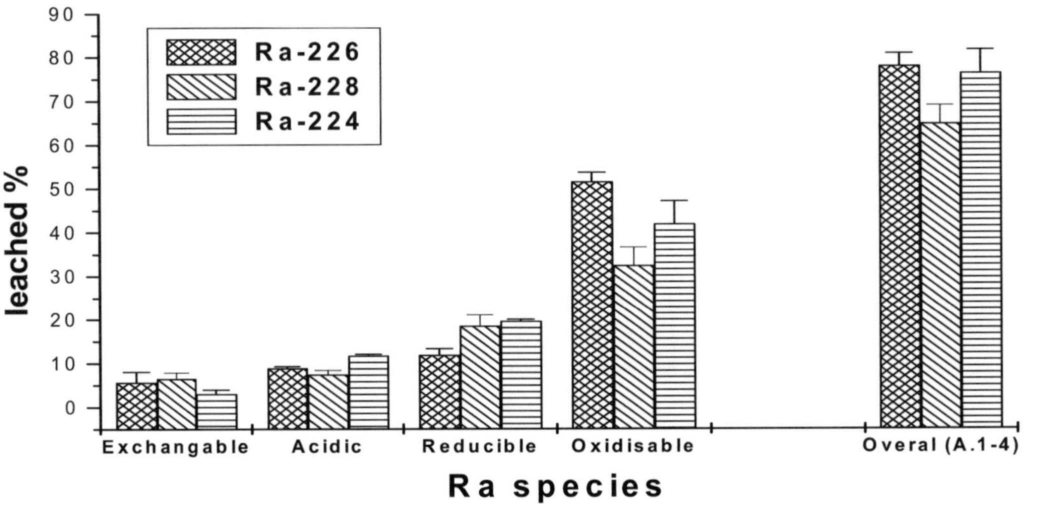

Figure 6. Distribution of radium species in T E NORM ludge using approach A

5.1.2. Approach 2

In approach 2, the TE-NORM waste was treated sequentially using different chemical leaching, through four leaching steps. The de-aerated and de-ionized H_2O (pH 6.7, 25 ± 1°C, 4 h), 1M CH_3COONH_4 (pH 6.8, 25 ± 1°C, 4 h) as water-soluble and exchangeable solutions for removal of Ra species are used. The results showed that the leached percentages (%) of radium isotopes are 10.6 ± 1.5, 9.7 ± 1.2, and 11.2 ± 0.8 % for Ra226, Ra228, and Ra224, respectively (step B.1).

In the second leaching process (B.2), the acidic radium species such as carbonates, and in addition, some iron and manganese oxides are removed. The leaching percentages (%) of radium species are 12.8 ± 2.8, 15.2 ± 0.5, and 16.5 ± 1.2% for Ra226, Ra228, and Ra224, respectively. The remaining waste was leached through two successive substeps. The solutions used are selective to the reducible radium species in the waste, such as manganese oxides, amorphous iron oxide, and moderately reducible phase (step B.3). The obtained removal percentages (%) of Ra226, Ra228, and Ra224 are 14.2 ± 1.2, 17.4 ± 3.1, and 19.0 ± 1.5%, respectively. Finally, the remaining waste was treated using oxidizing reagent solution, as a selective chemical agent to leach the oxidizable radium species in the waste (step B.4). The leached percentages (%) of the oxidizable Ra species are 53.3 ± 1.2, 48.4 ± 1.9, and 45.0 ± 2.3% for Ra226, Ra228, and Ra224, respectively, as shown in Figure 7.

From the data obtained from two leaching sequence mentioned above, it was found that using selective chemical solutions is more efficient when dealing with the different radium species present in the TE-NORM waste. Also, the data showed that the overall removal percentages (%) of all radium species are 90.9 ± 3.5, 86.7 ± 4.1, and 89.7± 6.2% for Ra226, Ra228, and Ra224, respectively. These values indicate that the amounts leached of the three radium isotopes by this approach are nearly the same. Figure 8 represents distribution of the actual removed (%)

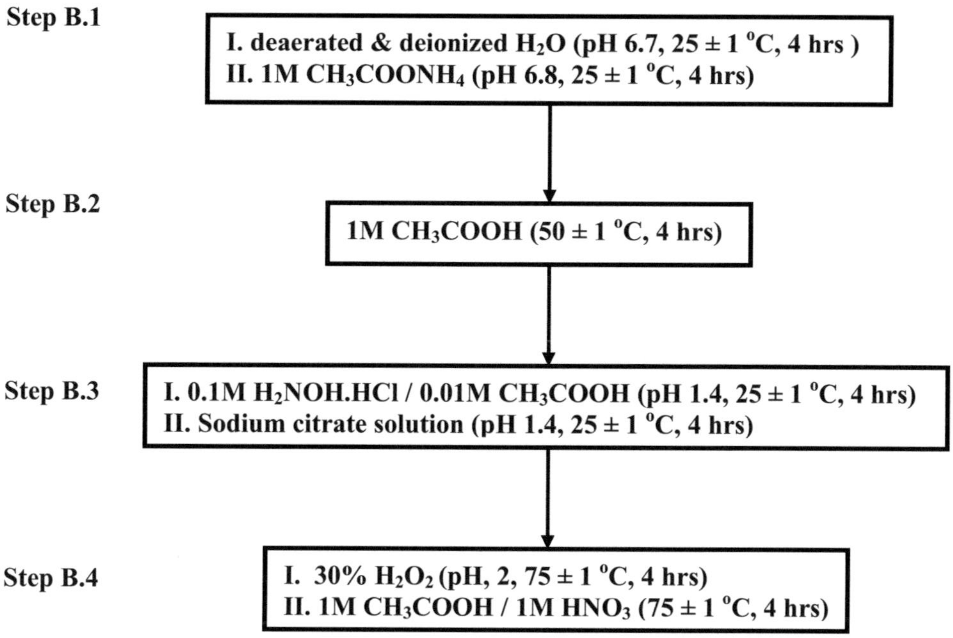

Figure 7. Sequential leaching of the radium content in TE-NORM was based on the individual extraction for each Ra species in the waste

toward different types of the radium species found in the treated TE-NORM waste. It is found that the high removal % of Ra226 is obtained for the radium oxidizable species. This is due to the high ability of the leaching solutions used in step (B.4) to remove the radium species from sludge waste. This conclusion confirms that the same behavior is obtained when using approach A. Therefore, the oxidizable Ra species is the main Ra fraction in waste. So, the net conclusion, the sequence of the different radium species present in the treated waste by leaching (%) can be ordered as: oxidizable > reducible > acidic > exchangeable as shown in Figure 8.

The overall removal (%) of the radium species using the both approaches (A and B) are illustrated in Figure 9. It is showed that the overall removal percentages of Ra226 and Ra224 are nearly the same when the waste is leached using approaches A and B. It is found that values of the overall removal % of Ra226 and Ra224 leached using solutions of approach A are 78% and 76%, respectively. On the other hand, it was found that the overall removal % using solutions of approach B is increased to ~90 % for Ra226 and Ra224. While the overall leached % of Ra228 is low comparable to Ra226 and Ra224 at the same leaching conditions, the obtained overall removal percentages of Ra228 are ~65% and 87.5% using solutions of the approaches A and B, respectively (Figure 9). The variation in the overall removal % between the leached Ra species from the TE-NORM sludge waste under the same leaching conditions is difficult to be explained. Finally, treatment of the sludge waste using solutions of approach B is more efficient compared to approach A, toward the overall removal percentages of Ra species.

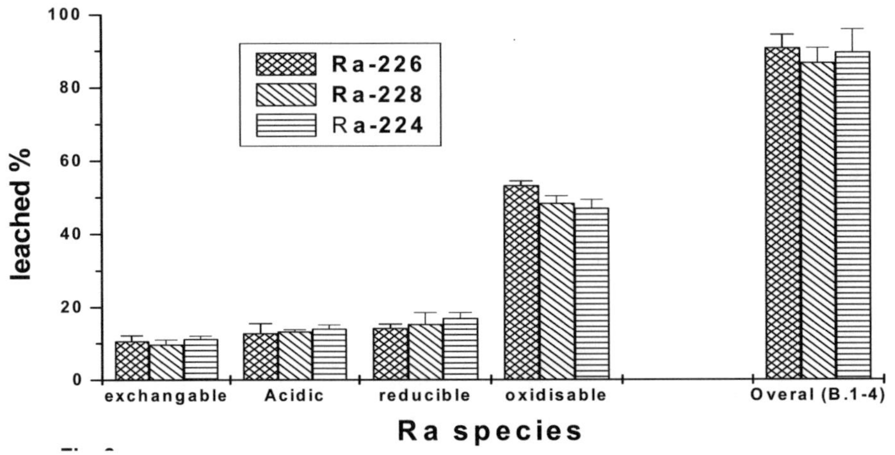

Figure 8. Distribution of radium species in TENORM s ludge u sin g approach B

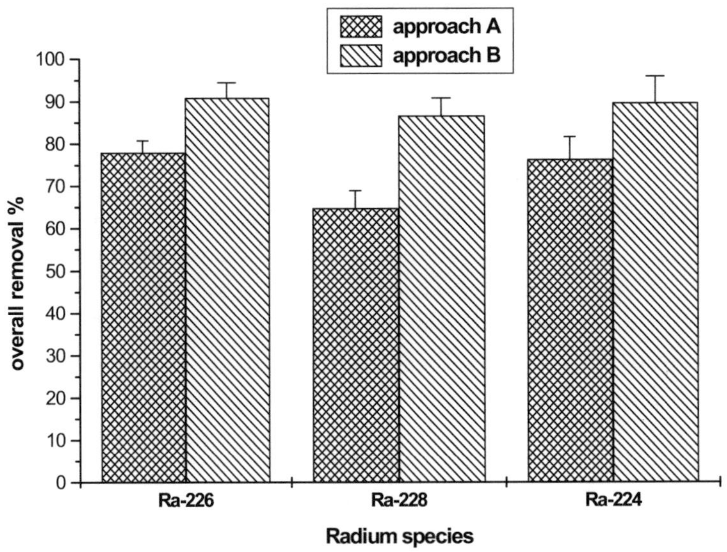

Figure 9. Overall removal percentages (%) of radium species leached from TENORM waste sludge using approaches A and B

The sequential chemical treatment could be the key point for environmental-friendly leaching for TE-NORM waste to select the suitable chemicals for the treatment processes [33].

The other alternative process for treatment of these wastes is leaching or solubilization of the different radionuclides. This is based on partial dissolution of the radionuclides using strong acids or by conversion of hardly or insoluble radionuclides forms to easily soluble salts. Within these merits, investigations were carried out to assess the direct leaching of radionuclides by HCl or HNO_3 or by treating the waste with carbonate solutions followed by leaching the

http://dx.doi.org/10.5772/61122

formed carbonates with dilute acid solution. The different conditions for the maximum removal of the radionuclides Pb210, Ra226, and Ra228 from the sludge and the scale wastes are given. From this table, it is clear that leaching with nitric acid produced better leaching efficiency for the removal of Pb210, Ra226, and Ra228 and then the use of hydrochloric acid. This is relating mainly to the oxidizing action of nitric acid. It is also clear that treatment with carbonate before leaching adds some benefits to the removal efficiency. This can be related to the possible conversion of the sulfate salts to the carbonate, which is easily leachable by dilute acids.

Leaching the carbonate treated sludge and scale wastes by high acid concentration can produce better removal for the different radionuclides, Table 8, yet use of strong acid is not recommended for its hazardous action. Therefore, and out of the different leaching systems studied, it can be recommended that the treatment of both the sludge and the scale wastes by 10% Na_2CO_3 followed by leaching with 1 M HNO_3 solution is recommended. This treatment will remove more than 70% of Pb210, Ra226, and Ra228 from scale waste and more than 55% of the same radionuclides from the sludge waste. It is also noted that the % removal of Ra226 is different than that of Ra228. This suggests that Ra226 is concentrated in different species of these wastes [63].

Method	Removal, %					
	Pb210		Ra226		Ra228	
	Sludge	Scale	Sludge	Scale	Sludge	Scale
HCl (7.5M)	36±2.1	56±2.8	29±1.5	19±1	22±1.2	34±1.7
10%Na_2CO_3/0.5M HCl	63±3.1	77±3.8	52±2.8	50±2.5	39±2	56±2.8
10%$(NH_4)_2CO_3$/0.5MHCl	21±1.1	63±3.1	23±1.1	32±1.8	44±2.2	37±1.8
25%$(NH_4)_2CO_3$/0.5MHCl	55±2.6	50±2.5	56±2.8	8.5±0.4	66±3.3	44±2.1
10%Na_2CO_3/1M HCl	66±3.5	76±3.7	54±2.6	53±2.6	47±2.7	85±4.3
$(NH_4)_2CO_3$/1M HCl	56±2.6	67±3.4	58±2.7	48±2.4	68±4.1	63±3.2
HNO_3 (7.5M)	37±2.2	53±2.5	35±2	16±0.8	21±1	37±1.8
10%Na_2CO_3/1M HNO_3	67±3.5	72±3.7	76±4.3	74±3.7	55±2.6	83±4.2
$(NH_4)_2CO_3$/1M HNO_3	65±4	71±3.6	55±2.9	42±2.1	47±2.3	48±2.4

Table 8. Comparison of maximum % removal of environmental interest radionuclides by different methods

Now, interesting study was done by our team using the solvent extraction technique for treatment of TE-NORM at scales on the interior of a pipe used in exploration of gas and oil industry, and several parameters were studied, such as the effect of contact time, organic extractants concentration, organic liquid/solid ratio, temperature, effect of different aliphatic and organic diluents. From the data available up till now, it can be concluded that kerosene as a diluent has a good efficiency on the E % of the radionuclides with the different organic extractants used. The extraction percent order with different types of organic extractants for ^{226}Ra, separation of ^{228}Ra, ^{238}U, ^{210}Pb, and ^{40}K at kerosene was found in the following order:

TOPO ≈ TBP > TBPO > DEHPA > TPPO > TPAsO for ^{226}Ra

TBP > TOPO > DEHPA > TBPO > TPPO > TPAsO for ^{228}Ra

TBP > DEHPA > TPPO > TBPO > TOPO > TPAsO for ^{238}U

TOPO > TBPO > TBP > DEHPA > TPAsO > TPPO for ^{210}Pb

TBP > DEHPA > TOPO > TBPO > TPAsO > TPPO for ^{40}K

Also, our team examined and evaluated two surfactants as extracting agent for the removal of radium species from TE-NORM sludge produced from petroleum industry. In this investigation, cationic and nonionic surfactants were used as extracting agents for the removal of radium radionuclides from the sludge waste. Two surfactants namely cetyltrimethylammonium bromide (CTAB) and Triton X-100 (TX100) were investigated as the extracting agents. Different parameters affecting the removal of both ^{226}Ra and ^{228}Ra by the two surfactants as well as their admixture were studied by the batch technique [64]. The influence of contact time on dissolution/desorption of radium radionuclides (^{226}Ra, ^{228}Ra) from TE-NORM sludge waste using TX100 and CTAB surfactants was investigated. Transport and mass transfers of radium isotopes from the sludge might be a key process responsible for reducing radium from the sludge. To achieve maximum radium species removal, a specific period of time is required. The obtained results are represented in Table 9. It is obvious that the removal efficiency of radium isotopes is increased as the shaking time was increased and reach maximum after 60 minutes. The highest removal efficiency for ^{226}Ra was obtained using CTAB surfactant, and using TX100 surfactant for ^{228}Ra. However, further increase in the time of experiment leads to decrease of the removal efficiency [64].

Time, min	TX100 solution		CTAB solution	
	^{226}Ra (R, %)	^{228}Ra (R, %)	^{226}Ra (R, %)	^{228}Ra (R, %)
15	16.0 ± 1.3	17.0 ± 1.4	20.4 ± 1.8	15.0 ± 1.3
30	22.0 ± 1.5	20.5 ± 1.8	22.7 ± 2.0	18.2 ± 1.6
60	25.0 ± 1.7	27.0 ± 2.1	26.0 ± 2.0	22.0 ± 1.3
120	15.7 ± 1.3	24.0 ± 2.2	23.7 ± 2.1	23.7 ± 2.0
240	8.2 ± 0.8	18.5 ± 1.6	4.2 ± 0.4	16.7 ± 1.5

Table 9. Effect of contact time on the removal efficiency (R, %) of ^{226}Ra and ^{228}Ra using 1% (w/v) surfactants solutions

Effect of surfactant concentration on the extraction of radium isotopes is regarded as an important parameter affecting the removal of radium isotopes from TE-NORM sludge waste. The removal efficiency of radium species for both surfactant solutions increased with increasing surfactant concentration up to 1%. At higher surfactants concentration, a slight decrease was observed in Table 10. The optimum concentrations are found to be 1% for both surfactants solutions. The effective removal of radium species from TE-NORM sludge can be explained by the increased solubility of radium species in the surfactant micelles. Generally, the change in the concentration of surfactant leads to change in its physical properties such as micelles formation and its solubilization effect for radium species or any contaminant (organic or inorganic species) present in TE-NORM sludge waste [65]. Therefore, the optimum surfactants concentrations are 1% for this treatment to avoid introduction of excess surfactants into sludge and avoid decrease in the radium removal %.

Concentration%	TX100 solution		CTAB solution	
	^{226}Ra (R, %)	^{228}Ra (R, %)	^{226}Ra (R, %)	^{228}Ra (R, %)
0.25	13.5 ± 1.4	15.0 ± 1.3	12.5 ± 0.9	14.4 ± 1.3
0.5	17.1 ± 1.1	19.1 ± 1.5	22.7 ± 1.7	20.0 ± 2.0
1.0	25.0 ± 1.7	27.0 ± 2.2	26.0 ± 2.0	22.0 ± 1.3
2.0	11.9 ± 1.1	20.5 ± 1.9	24.9 ± 1.7	23.0 ± 2.2
4.0	9.8 ± 0.9	19.0 ± 1.5	16.9 ± 1.4	14.8 ± 1.3
8.0	9.0 ± 0.8	13.3 ± 1.2	6.8 ± 0.6	8.8 ± 0.9

Table 10. Effect of surfactants concentrations on the removal efficiency (R, %) of ^{226}Ra and ^{228}Ra

The effect of temperature on the surfactants is not straightforward [66]. So that, temperature of surfactant solutions used for removal of radium species is an important parameter in surfactant-aided sludge washing process, and the experiments have been investigated with concentration of 1% TX100 and CTAB at 25–60°C. The results in Table 11 showed that the removal of Ra-isotopes are increased with increasing temperature and the removal of Ra species reach a maximum at 60°C using both surfactants solutions. The increase of Ra species removal efficiency is due to the properties of surfactants, where an increase in temperature generally results in an increase in the extent of solubility. The cloud point phenomenon occurs when a surfactant above its CMC causes the separation of the original solution into two phases when heated at a characteristic temperature called cloud point temperature. At this temperature, surfactant is no longer soluble in water and solution becomes hazy and cloudy. Above the cloud point, micelles formed from surfactant molecules act as an organic solvent in liquid–liquid extraction and the analytes are partitioned between the micelles and the aqueous phases [67]. It has been mentioned that the cloud point extraction procedure not only effectively solubilizes and concentrates pollutants but also appears to offer a means to further the concentrated surfactant-enhanced wash solutions that have been used in soil treatment processes [68]. About 25% of the radium species were initially removed from the TE-NORM sludge by solubilization in surfactants solution. About 55–60% removal was achieved upon the temperature raise to 60°C as shown in Table 11.

Synergism in surfactants may be defined as any situation where mixtures of surfactants have superior properties when compared to the properties of any of the single surfactant alone [69].

Temperature °C	TX100 solution		CTAB solution	
	^{226}Ra (R, %)	^{228}Ra (R, %)	^{226}Ra (R, %)	^{228}Ra (R, %)
25	25.0 ± 1.7	27.0 ± 2.2	26.0 ± 1.9	22.0 ± 1.3
35	30.0 ± 1.9	39.5 ± 2.2	28.6 ± 2.2	35.9 ± 2.8
45	53.0 ± 1.7	45.3 ± 2.0	43.5 ± 2.7	46.2 ± 3.4
60	58.6 ± 2.1	54.3 ± 3.3	49.8 ± 3.8	50.7 ± 4.2

Table 11. Effect of temperature on the removal efficiency (R, %) of ^{226}Ra and ^{228}Ra using surfactants solutions

There is usually a synergy effect for the CMC of surfactant mixtures (mixture of nonionic and ionic surfactant) [69]. Mixture of TX100 and CTAB surfactants showed synergistic interactions, which can be manifested as enhanced surface properties, spreading, and many other phenomena, as shown in Figure 5. The synergistic behavior of mixed surfactant systems can be exploited to reduce the total amount of surfactant used in a particular application resulting in the reduction of cost [70]. It was observed that the removal values of radium isotopes of mixed systems of both surfactants are higher than their corresponding values without mixing, which indicate synergistic interaction in mixed CTAB-TX100 as a chemical extraction system. Removal of 84% and 80% for ^{226}Ra and ^{228}Ra, respectively, are obtained using synergistic effect of 1% aqueous solution containing 1:1 of the two surfactants investigated. In other words, mixed micelle formation in aqueous solution can be greater than that of the individual surfactant, and explained by non-ideal solution theory [70]. Also, it was observed that combined extraction of cationic and nonionic surfactants was effective in removal of both ^{226}Ra and ^{228}Ra. Experiments indicated that removal efficiency was optimized (80–84%) when a mixture of 1% CTAB and 1% TX100 was employed at the ratio 1:1. The theoretical justification for this surfactants solution is based upon two hypotheses, first that surfactant micelles may sequester radium radionuclides which are sorbed to the TE-NORM sludge waste, and second that the surfactant micelles may increase the concentration of radium radionuclides in the aqueous phase. The developed chemical treatment process would enable to design an appropriate TE-NORM sludge washing strategy.

Glossary

Brief explanations of few terms are mentioned in the state-of–the-art in this short glossary.

NORM: Naturally occurring radioactive material.

TE-NORM: Technologically enhanced naturally occurring radioactive materials.

by-product: Any product from an (industrial) process that is not the intended primary product, but inevitable given the process implemented.

http://dx.doi.org/10.5772/61122

Product: The intended output from an (industrial) process.

REE: Rare earth elements (or lanthanides), comprising 16 chemical elements, including those with atomic numbers 57 (lanthanum) through 71 (lutetium), as well as yttrium (atomic number 39).

Waste: Any material that is:

(a) The unwanted/substandard output from an (industrial) process that cannot be utilized under given circumstances (technological/economic perspective);

(b) Declared as unwanted and/or unusable (regulatory perspective).

Scale: Solid deposit of low-solubility sulfates or carbonates on the inside of components of gas and oil production installations.

Sludge: Mixture of organic and mineral solids in water and liquid hydrocarbons separated from oil or gas at production facilities.

Radioactive material: Material designated in national law or by a regulatory body as being subject to regulatory control because of its radioactivity.

Half-life: For a radionuclide, the time required for the activity to decrease, by a radioactive decay process, by half. i.e., lose 50% of its activity.

Secular equilibrium: is a steady-state condition in which the rate of formation of the radioactive daughter products is just equal to the rate of decay; that is, the activity of radioactive parent and daughter are the same.

Separator: A pressure vessel used for separating well fluids produced from oil and gas wells into gaseous and liquid components.

Industrial process: This term is used very broadly in the present report to denote any human activity involving the application of technology, for example, the mining, processing, and drinking water treatment industries.

EPA: Environmental Protection Agency

UNSCEAR: United Nations Scientific Committee on the Effects of Atomic Radiation

TBP: Tri butyl phosphate

DEHPA: Bis (2-ethylhexyl phosphoric acid)

TOPO: Tri octyl phosphine oxide

TBPO: Tri Butyl phosphine oxide

TPAsO: Tri phenyl arsine oxide

TPPO: Tri phenyl phosphine oxide

Author details

N.S. Awwad[1,3*], M.F. Attallah[3], E.M. El-Afifi[3], H.A. Ibrahium[2] and H.F. Aly[3]

*Address all correspondence to: nsawwad20@yahoo.com

1 Chemistry Department, Faculty of Science, King Khalid University, Abha, KSA

2 Biology Department, Faculty of Science for Girls, King Khalid University, Abha, KSA

3 Hot Laboratories and Waste Management Center, Atomic Energy Authority, Cairo, Egypt

References

[1] OGP Report No. 412, Guidelines for the management of Naturally Occurring Radioactive Material (NORM) in the oil & gas industry, September 2008, www.ogp.org.uk. ©Copyright 2008, OGP.

[2] UNSCEAR (1977) United ionising radiations: sources and biological effects of atomic radiation. Report to the General Assembly. Vienna, New York.

[3] Al-Masri MS, Shwiekani R. Radon distribution in natural gas processing facilities and workplace air environment. J Environ Radioact 2008;99:574–580.

[4] Dickson BL. Radium in groundwater. In: The Environmental Behavior of Radium. Reports Series, Vol. 310, No 1. IAEA, Vienna, 1990, pp.335–372.5.

[5] Jonkers G, Hartog FA, Knappen AAI, Lance PFJ. Characterization of NORM in the oil and gas production (E&P) industry. In: Proc NORM I, Amsterdam, 1997, pp.23–47.

[6] Snavely ES. Radionuclides in produced water. Report to the American Petroleum Institute. Publication no 5404. API, Washington, DC, 1989, pp.1–86.

[7] White GJ. Naturally occurring radioactive materials (NORM) in oil and gas industry, equipment and wastes: a literature review. Report DOE/ID/01570-T158.Bartlesville, 1992.

[8] Hamlat MS, Djeffal S, Kadi H. Assessment of radiation exposures from naturally occurring radioactive materials in the oil and gas industry. Appl Radiat Isot 2001;55:141–146.

[9] Guidelines for naturally occurring radioactive materials Australian Petroleum Production & Exploration Associated Ltd. Report ABN 44000292773, March 2002, Canberra.

[10] Vegueria JSF, Godoy JM, Miekeley N. Environmental impact studies of barium and radium discharges by produced waters from the "Bacia deCampos" oil field offshore platforms, Brazil. J Environ Radioact 2001;62:23–38.

[11] Testa C, Desideri C, Meli MA. Radiation protection and radioactive scales in oil and gas production. Health Phys 1994;71:34–38.

[12] Shawky S, Amer H, Nada AA, Abd El-Maksoud TM, Ibrahiem NM. Characteristics of NORM in the oil industry from Eastern and Western deserts of Egypt. Appl Radiat Isot 2001;55:135–139.

[13] Norwegian Radiation Protection Authority Natural radioactivity in produced water from the Norwegian oil and gas industry in 2003. Report no 2. NRPA, Ǿsteras, 2005.

[14] Eriksen DO, Sidhu R, Strälberg E. Radionuclides in produced water from Norwegian oil and gas installations-concentrations and bioavailability. Czechoslovak J Phys 2006;56:D43–D48.

[15] Al-Masri MS. Spatial and monthly variations of radium isotopes in produced water during oil production. Appl Radiat Isot 2006;64:615–623.

[16] United Kingdom Off-Shore Operations Association UK North Sea oil and gas industry; environmental inputs, impacts and issues. A report prepared by Environmental and Resource Technology Ltd, London, 1992.

[17] Stephenson MT, Supernow IR. Offshore Operators Committee 44 Platform study radionuclide analysis results. Offshore Operation Committee Report, New Orleans, Louisiana, 1990.

[18] Swan C, Matthews J, Ericksen R, Kuszmaul J Evaluation of radionuclides of uranium, thorium, and radium associated with produced water fluids, precipitates and sludge from oil, gas and oilfield brine injections wells in Mississippi. US DOE Report; 2004, DE-FG26-02NT 15227.

[19] Zieliński RA, Budahn JR. Mode of occurrence and environmental mobility of oil-field radioactive material at US Geological Survey research site B. Appl Geochem 2007;22:2125–37.

[20] Godoy MJ, da Cruz RP. 226Ra and 228Ra in scale and sludge samples and their correlation with the chemical composition. J Environ Radioact 2003;70:199–206.

[21] Gazineu MHP, de Araujo AA, Brandao YB, Hazin CA, Godoy JM. Radioactivity concentration in liquid and solid phases of scale and sludge generated in the petroleum industry. J Environ Radioact 2005;81:47–54.

[22] Gazineu MHP, Hazin CA. Radium and potassium-40 in solid wastes from the oil industry. Appl Radiat Isot 2008;60:90–94.

[23] El Afifi EM, Awwad NS. Characterization of the TENORM waste associated with oil and natural gas production in Abu Rudeis. Egypt. J Environ Radioact 2005;82:7–19.

[24] Kadyrzhanov KK, Tuleushev AZ, Marabaev ZN. Radioactive components of scales at the inner surface of pipes in oil fields of Kazakhstan. J Radioanal Nucl Chem 2005;264:413–6.

[25] Omar M, Ali HM, Abu MP. Distribution of radium in oil and gas industry wastes from Malaysia. Appl.Radiat Isot 2004;60:779–782.

[26] Lysebo J, Birovliev A, Strand T. NORM in oil production – occupational doses and environmental aspects. In: Proc of the 11th Congress of the Nordic Radiation Protection Society, 26–30 August 1996, Reykjavik, p.137.

[27] Al-Saleh FS, Al-Harshan GA. Measurements of radiation level in petroleum products and wastes in Ryad City refinery. J Environ Radioact 2008;99:1026–1031.

[28] Heaton B, Lambley JG. TENORM in the oil and gas industry. Appl Radiat Isot 1995;46:577–581.

[29] Exploration and Production Forum. Low specific activity scale origin treatment and disposal. Report no 6.6/127. London, 1987;pp.25–38.

[30] Scot ML. Naturally occurring radioactive materials in non-nuclear industry. In: Proc of the 2nd Int. Symp. on the Treatment of Naturally Occurring Radioactive Materials NORM II, 10–13 November 1998, Klefeld, Germany, pp.163–167.

[31] White GJ, Rod SA. Radon emanation from NORM-contaminated pipe scale and soil at petroleum industry sites. J Environ Radioact 2001;54:401–413.

[32] Oil and gas production wastes. http://www.epa.gov/rpdweb00/ tenorm/ oilandgas.html.

[33] El Afifi EM, Awwad NS, Hilal MA. Sequential chemical treatment of radium species in TENORM waste sludge produced from oil and natural gas production, J Hazard Mater 2009;161:907–912.

[34] El Afifi EM. Radiochemical studies related to environmental radioactivities. Ph.D. thesis (Chemistry), Faculty of Science, Ain Shams University, Cairo, Egypt, 2001, p. 98.

[35] Shawky S, Amer H, Nada AA, Abdel Maksoud TM, Ibrahim NM. Characteristics of NORM in the oil industry from eastern and western deserts of Egypt. Appl Radiat Isot 2001;55:135–9.

[36] Holland B. Experience with operations involving NORM in the UK and some other regions. Australian Nuclear Science and Technology Organization, Lucas heights, Sydney, March 16–20, 1998.

[37] White GJ, Rood AS. Radon emanation from NORM contaminated pipe scale and soil at petroleum industry sites. J Environ Radioact 2001;54:401–413.

[38] Hamlat MS, Gjeffal S, Kadi H. Assessment of radiation exposures from naturally occurring radioactive materials in the oil and gas industry. Appl Radiat Isot 2001;55:141–146.

[39] Heaton B, Lambley J. TENORM in the oil, gas and mineral mining industry. Appl Radiat Isot 1995;46:577–581.

[40] Exploration & Production Forum. Low specific activity scale origin treatment and disposal. Report no. 6.6/127, Old Burlington Street, London WlX 1LB, 1987, pp.25–28.

[41] Hilal MA, Awwad NS, El Afifi EM, Aly HF. TE-NORM in oil-separator tanks, assessment and treatment investigations. Arab J Nuc Sci Applic 2005;39(2):109–124.

[42] Environmental, Health, and Safety Guidelines Offshore Oil and Gas Development, APRIL 30, 2007.

[43] UNSCEAR United Nations Committee on the Effect of Atomic Radiation: Sources and NCRP. Exposure of the population in the United Sates and Canada from natural background radiation. NCRP report no.94. National Council on Radiation Protection and Measurement, Bethesda, Maryland, 1994.

[44] IAEA. International Atomic Energy Agency. Report of analysis on Determination of thorium and uranium naturally occurring radioisotopes in IAEA reference materials. IAEA laboratories (Chemistry Unit-01-10), Seibersdorf Austria, pp.1–5.

[45] Eylander JGR, Lancee PFJ, Hartog F, Knaepen WAL, Frigo DM. Proc 2nd Int Symp Treatment NORM, p.83, Krefeld, Germany, Nov. 10–13, 1998.

[46] Quada U, Kluth T. Proc 2nd Int Symp Treatment NORM, p.95, Krefeld, Germany, Nov. 10–13, 1998.

[47] Stage G. Proc 2nd Int Symp Treatment NORM, p.95, Krefeld, Germany, Nov. 10–13, 1998.

[48] Smith GE, Fitzgibbon T, Karp S. Proc Exploration Production Environ Conf p.231, Houston, Texas, USA, March 27–29, 1995.

[49] Daifullah AAM, Awwad NS. Proc 5th Int Sci Conf 25–27 March, pp.57–69, 2003.

[50] Hassan SSM, Awwad NS, Aboterika AHA. Removal of mercury(II) from wastewater using camel bone charcoal, J Hazard Mater 2008;154:992–997.

[51] Phillip E, Carter H. Health Physics Society, Albuquerque, NM, 24 January, 1999.

[52] Exploration and Production Forum. Low specific activity scale origin treatment and disposal. Report no. 6.6/127, Old Burlington Street, London, WI1X 1LB, UK, 1987, pp. 25–28.

[53] IAEA, International Atomic Energy Agency, International basic safety standards for the protection against ionizing radiation and for the safety of radiation sources. GOV / 2715 / Vienna, 1994.

[54] Rood AS, Kendrick DT. Measurement of Rn222 flux, Rn222 emanation and Ra-226 concentration from injection well pipe scale, NORM/NARM: Regulation and risk assessment, Proceedings of the 29th midyear topical meeting of the Health Physics Society, Scottsdale, Arizona, USA, January 7–10, 1996, pp.139–144.

[55] Smith GE, Fitzgibbon T, Karp S. Economic impact of potential NORM regulation, SPE/EPA exploration and production environmental conference, Houston, Texas, USA, March 27–29, 1995, pp.181–194.

[56] Eylander JGR, Lancee PFJ, Hartog FA, Knaepen WAI, Frigo DM. Further developments in the in-situ removal of TENORM from exploration and production facilities. Proc 2nd Int Symp Treatment NORM, Krefeld, Germany, November 10–13, 1998, pp. 83–89.

[57] Shehata FA, El Afifi EM, Aly HF. Fractionation and leaching of naturally occurring radioactive waste produced from oil production, Pro Conf Radioactive Waste Management Environ Remediation Nagoya, Japan, September 26–30, 1999, CD-ROM.

[58] El Afifi EM, El-Reefy SA, Aly HF. Treatment of solid waste containing Ra-226. Arab J Nucl Sci Appl 2006;39:35–47.

[59] Benes P. Speciation procedures. In: The environmental behavior of radium, Technical reports series No. 310, Vol. 1, Part 3, Ch. 10, International Atomic Energy Agency (IAEA), Vienna, 1990, pp.273–299.

[60] Florence TM, Batley GE. Chemical speciation in natural waters. CRC, Crit Rev Anal Chem 1980;9:219–296.

[61] Sam AK, Ahmed MMO, El Khangi FA, El Nigumi YO, Holm E. Radiological and chemical assessment of Uro and Kurun rock phosphates. J Environ Radioact 1999;42:65–75.

[62] Tessier A, Campbell PGC, Bisson M. Sequential extraction procedure for the speciation of particulate trace metals. Anal Chem 1979;51:844–851.

[63] Attallah MF, El Afifi EM, Awwad NS, Aly HF. Comparative study on the radioactivity of TE-NORM in different components of oil separator tanks Radiochimica Acta 2102;101: 57– 65.

[64] Attallah MF, Hamed MH, El Afifi EM, Aly HF. Removal of Ra-226 and Ra-228 from TENORM sludge waste using surfactants solutions. Accepted for publication in Journal of Environmental Radioactivity, 2014.

[65] Crook EH, Fordyce DB, Trebbi GF. Molecular weight distribution of nonionic surfactants. I. Surface and interfacial tension of normal distribution and homogeneous p, t-octylphenoxyethoxyethanols (OPE'S). J Phys Chem 1963;67:1987.

[66] Singh HN, Swarup S, Saleem SM. Effect of electrolytes on the micellization of ionic surfactants in n-alkanol-water mixtures. J Colloid Inter Sci 1979;68:128.

http://dx.doi.org/10.5772/61122

[67] Schick MJ. Nonionic Surfactants: Physical Chemistry, CRC Press; 2nd edition, 1987.

[68] Abdul AS, Gibson TL. Laboratory studies of surfactant enhanced washing of polychlorinated biphenyl from sandy material. Environ Sci Technol 1991;25:665.

[69] Holmberg K, Jönsson B, Kronberg B, Lindman B. Surfactants and Polymers in Aqueous solution, 2nd Ed., Wiley, Chichester, 2003.

[70] Stalikas C.D. Micelle mediated extraction as a tool for separation and preconcentration in metal analysis, Trends Anal Chem 2002;21:343.

Section 3

Modeling and Simulation

Chapter 6

Simulation and Optimization of Multi-period Steam Cracking Process

Lei Zhang, Hangzhou Wang, Tong Qiu and Bingzhen Chen

Additional information is available at the end of the chapter

http://dx.doi.org/10.5772/60558

Abstract

Hydrocarbon steam cracking is the most important process for producing industrial chemicals such as olefin and aromatics. Steam cracking modelling and optimization is an effective way for increasing production and saving energy. In this chapter, multi-scale modelling and elementary reaction networks are established and used in the modelling and optimization of steam cracking. However, the large scale of the optimization model makes it difficult to obtain a solution. Thus, a surrogate coke thickness model for long-term steam cracking is proposed in this chapter to remove the connection between different periods of steam cracking process. By so doing, a parallel simulation can be used to accelerate optimization. An industrial case study showed optimization time to be significantly reduced from 17.78 hours to 2.08 hours using multi-period optimization with parallel simulation and the surrogate coke thickness model. It has been shown that a 0.62% increase in ethylene yield can be obtained via operating condition optimization, which demonstrates the effectiveness of the multi-scale steam cracking model and multi-period optimization with parallel simulation.

Keywords: Steam cracking, multi-scale modelling, elementary reaction network, parallel simulation, surrogate model

1. Introduction

Hydrocarbon steam cracking is the most important process for producing industrial chemicals such as olefin and aromatics. Mathematical models for steam cracking simulation have been

INTECH
open science | open minds

© 2015 The Author(s). Licensee InTech. This chapter is distributed under the terms of the Creative Commons Attribution License (http://creativecommons.org/licenses/by/3.0), which permits unrestricted use, distribution, and reproduction in any medium, provided the original work is properly cited.

studied for several decades and various models have been developed such as SPYRO [1] and COILSIM [2]. Meanwhile, the steam cracking model can be applied to the steam furnace design and optimization [3]. As various feedstock and furnaces are used in steam cracking, a more accurate and robust model is needed.

A mathematical model is a collection of relations among variables representing certain properties of a system, using equations and inequalities. By solving a model, one can predict the values of some properties of the system, given the values of some other properties, or predict the distributions of the values of some properties in particular domains, given certain boundary conditions. To meet the scientific and engineering demands in terms of revealing the characteristics of a system in-depth, a mathematical model may involve relations of properties at different scales of the system [4], referred to as a multi-scale model. The multi-scale model often consists of nonlinear equations and differential equations, and is not easy to solve. The information communication between scales is the key factor in multi-scale models [5]. In recent years, multi-scale modelling has been applied in all fields of chemical engineering such as thermodynamics, reaction engineering, polymer materials and CFD (computational fluid dynamics), among others. Multi-scale modelling is used in steam cracking models to reveal the nature of steam cracking and to establish a more accurate and robust model.

The kernel part of the steam cracking model is the reaction network. Researchers have developed various reaction models to describe the steam cracking process. There are three different types of reaction models: empirical models, molecular models and elementary reaction models. Empirical models use a large database of experimental results to regress a number of empirical correlations for the yields of the main products as a function of a number of easily measurable process variables [6]. Empirical reaction models need a large number of experiment data to render a regression. Once the feedstock or the furnaces change, the reaction model has to be re-established to obtain accurate simulation results. On the basis of empirical reaction models, molecular reaction models have been developed and frequently used [7-9], e.g., the Kumar reaction model [10]; 22 molecular reactions are involved in Kumar's reaction model: one primary reaction and 21 secondary reactions. The primary reaction is shown as eq. (1). The selective coefficients of the primary reaction (a_1~a_{10}) have been regressed from experimental data. If the feedstock or operating conditions change, the selective coefficients of the primary reaction should be regressed again.

$$\begin{aligned}&\text{naphtha} \rightarrow a_1\text{H}_2 + a_2\text{CH}_4 + a_3\text{C}_2\text{H}_4 + a_4\text{C}_2\text{H}_6 + a_5\text{C}_3\text{H}_6 + \\ &+ a_6\text{C}_3\text{H}_8 + a_7\text{C}_4\text{H}_{10} + a_8\text{C}_4\text{H}_8 + a_9\text{C}_4\text{H}_6 + a_{10}\text{C}_4\text{'s}\end{aligned} \tag{1}$$

Nowadays, elementary reaction models have been widely used to develop a more reliable and robust mathematical model. Since the pioneering work of Rice [11-13] there has been a general consensus about the elementary reaction mechanism. As the mathematical difficulties encountered for solving the detailed kinetic models can be overcome by the development of stiff solvers [14], detailed elementary reaction networks are widely used to accurately describe chemical processes in a wide range of process conditions and feedstock. Sundaram et al. [15]

established a radical reaction model for the pyrolysis of simple paraffins, olefins and their mixtures. Scharfe et al. [16] established a radical reaction model for n-hexane pyrolysis. Joo et al. [17] established a radical reaction model for industrial naphtha cracking furnaces.

However, it has remained problematic to generate thousands of elementary reactions and to determine the reaction rate constants of these reactions until an automatic reaction network generation technique was studied. Today, computers are used not only to solve the simulation numerically, but also to generate the elementary reaction network, construct the model and calculate the kinetic parameters [2]. Many research groups have developed computer tools for automatically generating these mechanisms [18-23] such as RMG [24] and CRACKSIM [25]. The elementary reaction model is expected to be used in furnace design and operating condition optimization. However, the huge number of reactions in the elementary reaction network usually requires significant CPU time during simulation. Thus, the steam cracking simulation model has to be reduced before it can be used in operating condition optimization, in order to make real-time optimization (RTO) realistic.

There are multiple methods for creating model reduction. In terms of chemistry, quasi-steady-state approximation (QSSA) [26], reaction rate analysis [27] and reaction path analysis [28], among others, can be used to reduce a reaction model. In terms of mathematics, principle component analysis [29] and sensitivity analysis [30] can also be used. On the other hand, data driven methods can also be used in model reduction, for example, the black box model, a neural network [31] and PCA based ROMs [32]. Several assumptions can also be applied to help retain the mechanism at a manageable size. The most important assumption is the μ radical hypothesis, which assumes that bimolecular reactions can be neglected for radicals with more than five carbon atoms [33]. The latter are also referred to as μ radicals. Thus, the QSSA method can be applied to remove μ radicals from the reaction network.

A multi-scale model for the steam cracking process is established in this chapter. The multi-period steam cracking process is also studied in the context of the established multi-scale model. Coking is an unavoidable factor during the multi-period steam cracking process. Coking increases pressure drops in the reaction tube, decreases the coefficient of heat transfer between the furnace and tube, and most importantly, increases the outer-wall temperature of the tubes. If one of the tubes in the furnace reaches the maximum allowance temperature of the tube material, the furnace must be shut down to execute a decoking process, otherwise the tube will be destroyed. The operation conditions are generally maintained constant during the long-term steam cracking process. Thus, dynamic operating conditions need to be optimized using a detailed steam cracking model in order to achieve a higher profit. Abel et al. [34] used the SQP method to solve a real-time optimization problem in the olefins production process. Tarafder et al. [35] proposed a multi-objective optimization problem in the operation and design of a styrene manufacturing process. Li et al. [36] applied an artificial neural network (ANN) hybrid model in the operation optimization of a naphtha industrial cracking furnace. Gao et al. [3] used a new parallel hybrid algorithm combining NSGA-II with SQP on multi-objective optimization for the periodic operation of the naphtha pyrolysis process. However, due to the complexity of the elementary reaction model, the researchers did not use this reaction model in the optimization problem. In this chapter, an elementary reaction model is applied to the operating condition optimization problem to obtain a more reliable optimization result. Based on this, a surrogate coke thickness model is proposed to make multi-period optimization with parallel simulation possible.

The general idea of this chapter is outlined in Figure 1. The first step in the conceptual development of a detailed molecule-based model for a complex feedstock is to determine an accurate molecular representation of the feedstock. Then, a multi-scale steam cracking model is established following the feedstock prediction. Finally, operating condition optimization of multi-period steam cracking is carried out using the established multi-scale model.

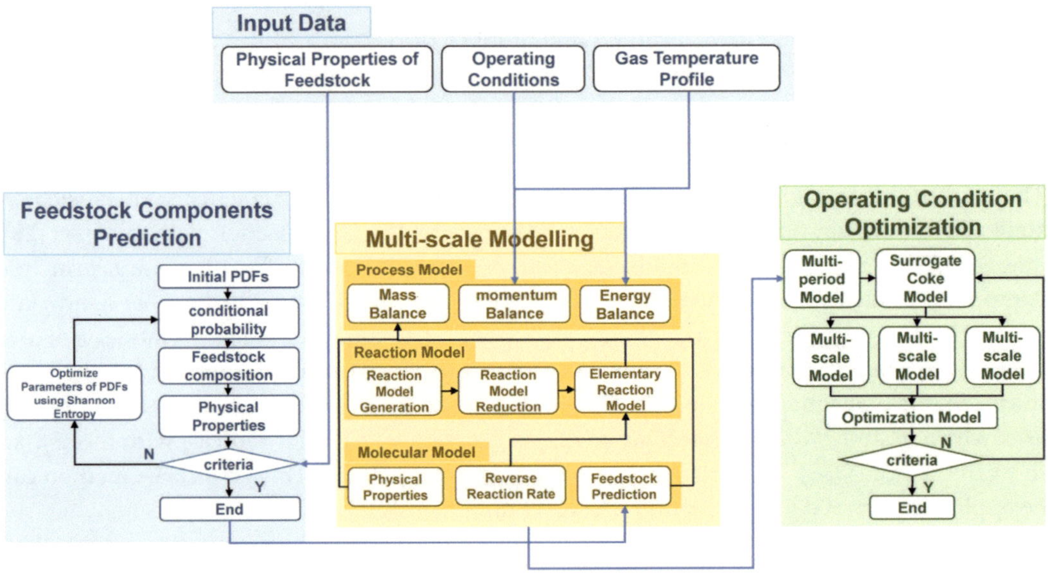

Figure 1. Diagram of simulation and optimization of the steam cracking process methodology

This chapter is structured as follows. Section 2 offers a detailed discussion of the establishment of the multi-scale steam cracking model; a case study for naphtha steam cracking simulation is presented. Section 3 provides the operating condition optimization model; surrogate coke thickness model and parallel simulation are used to accelerate the computing of the optimization model. A case study of operating condition optimization is presented and the results are presented and discussed. Finally, section 4 offers conclusions from this study.

2. Steam cracking model

2.1. Multi-scale model for steam cracking

A multi-scale model is proposed in this chapter to reveal the nature of the steam cracking process and to generate a reliable and robust model for accurately predicting the yields of

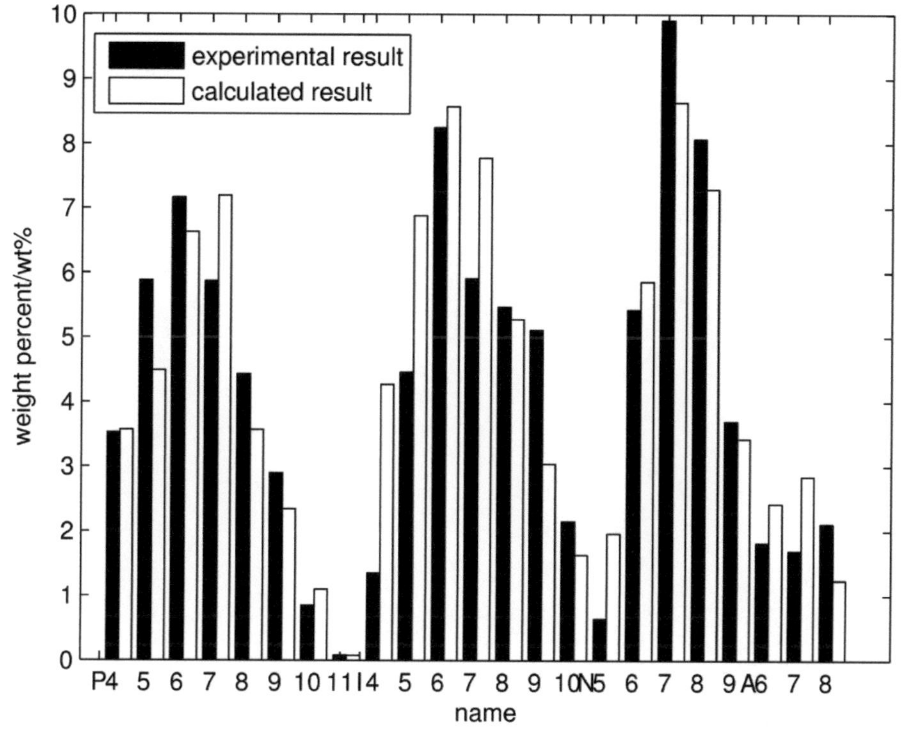

Figure 2. Comparison of experimental and calculated feedstock component mole fractions [39].

products. The multi-scale model is built up from the molecular level, to the reaction level to the process level. The establishment of the multi-scale model is discussed in detail in the following sections.

a. Feedstock composition prediction

Conventional analytical techniques are generally incapable of directly measuring the identities of all the molecules in complex feedstock, especially for the high carbon number range; however, this applies only to indirect characteristics [37]. Here, the Shannon entropy method [38] and probability density function were used to predict detailed feedstock composition, based on the analytical data. A detailed model for feedstock composition prediction can be found in [39]. The objective function of this NLP (non-linear programming) problem is shown in eq. (2), where *S(x)* is Shannon entropy and x_i is the mole fraction of component *i*.

$$\max S(x) = -\sum_{i=1}^{n} x_i \ln x_i \tag{2}$$

Figure 2 shows the predicted feedstock composition using Shannon entropy, compared with the experimental data. The predicted results show the effectiveness of the Shannon entropy theory in obtaining the missing information for models of the steam cracking process.

b. Process-level model

A process-level model consists of mass balance equations, momentum balance equation and energy balance equation (eqs. (3-5)) [40, 41]. For the simulation of smooth tubular reactor types, the use of a one-dimensional reactor plug-flow model is generally recognized as providing a sufficient degree of accuracy, as all radical profiles are wiped out due to the high turbulence corresponding to Reynolds numbers of over 250 000. The plug-flow reactor model implicitly assumes that there is no mixing in the axial direction, but rather, perfect mixing in the transverse direction [2].

$$\frac{dN_m}{dL} = \frac{S}{V}\sum_i v_{i,m} r_i = f_N(T,P,N_m) \tag{3}$$

$$\frac{dP}{dL} = -\frac{f \cdot E(L) \cdot G^2}{5.07 \times \rho \cdot D_i \times 10^4} = f_P(T,P,N_m) \tag{4}$$

$$\frac{dT}{dL} = \frac{k\pi D_o (T_w - T) - \sum_m \Delta H^0_{fm} \cdot \frac{dN_m}{dL}}{\sum_m c_{pm} N_m} = f_T(T,P,N_m) \tag{5}$$

In eq. (3), N_m is the concentration of species m and L is the length of the reactor tube. S is the flow area, V is the volume flow rate, μ_{im} is the stoichiometric coefficient of reaction i and r_i is the reaction rate of reaction i. In eq. (4), P is the pressure of the mass flow, f is the Fanning friction factor, $E(L)=L_e/L$, L_e is the equivalent length of the reactor tube, G is the mass flow rate, D_i is the internal diameter of the tube and ρ is the density of the gas mixture. In eq. (5), T is the temperature of the mass flow, D_o is the outer diameter of the tube, k is the overall heat transfer coefficient, T_w is the outer wall temperature of the tube, $\Delta H_{fm}{}^0$ is the standard heat of formation of species m and c_{pm} is the specific heat of species m.

It should be noted that the measuring point of the COT (coil-outlet temperature) of an industrial furnace is usually on the outer wall of the tube. Thus, the measured COT has a temperature difference from the outlet temperature of the gas mixture. Eq. (6) has been derived from heat balance equations in order to adjust to the temperature difference; the results show that there exists a 15-20K temperature difference, which agrees well with what has previously been reported.

$$\Delta COT = \frac{\lambda_{isolation}\left(COT - T_{isolation\_o}\right)}{\ln\frac{r_o + d_{isolation}}{r_o}} \left(\ln\frac{r_i}{r_i - d_{coke}} + \frac{\ln\frac{r_o}{r_i}}{\lambda_{tube}} + \frac{1}{\alpha_{mixture}(r_i - d_{coke})} \right) \tag{6}$$

http://dx.doi.org/10.5772/60558

In eq. (6), $\lambda_{isolation}$ is the heat transfer coefficient of the isolation layer. $T_{isolation\_o}$ is the outer wall temperature of the isolation layer, r_o is the external diameter of the tube, r_i is the interior diameter of the tube, $d_{isolation}$ is the thickness of isolation layer, d_{coke} is the thickness of coke and $\alpha_{mixture}$ is the heat transfer coefficient of the gas mixture in the tube.

c. Reaction-level model

The reaction model is the most important part of the steam cracking model. Many researchers have conducted in-depth studies on the elementary reaction model. An elementary reaction model can contain thousands of reactions and hundreds of species. The reaction model can be extremely hard to solve due to its stiffness. An accurate and robust elementary reaction model is developed in this chapter, and a Gear algorithm is used to solve the stiff ODEs. Generally, a detailed reaction network is generated by allowing the feedstock components to react according to different reaction families. The reaction families can be summarized as follows: (1) initiation reaction and termination reaction: $R_1-R_2 \leftrightarrow R_1\bullet+R_2\bullet$; (2) hydrogen abstraction reaction: $R_1-H+R_2 \leftrightarrow R_1\bullet+R_2-H$; (3) radical addition and β-scission reaction: $R_1\bullet+R_2=R_3 \leftrightarrow R_1-R_2-R_3\bullet$.

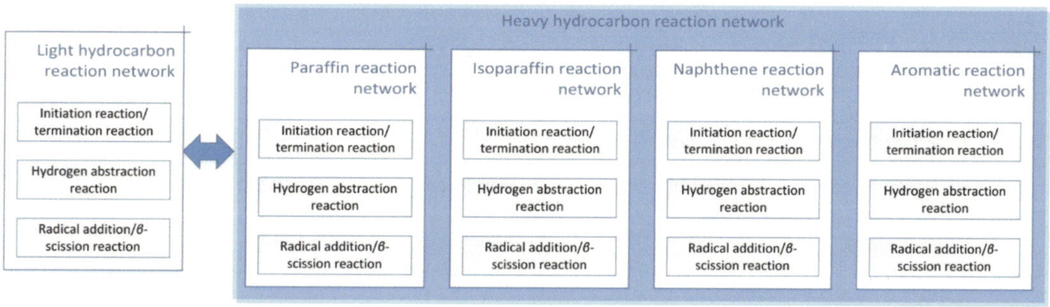

Figure 3. Sub-models in the generation of hydrocarbon steam cracking reaction networks.

The hydrocarbon steam cracking elementary reaction network can be separated into several sub-models. based on the composition of feedstock as shown in Figure 3. The sub-models are generated separately, based on the reaction families described above.

An elementary reaction model for light hydrocarbon can be found in much of the literature and databases, and it is more accurate than the automatic generated elementary reaction model. Thus, the elementary model for steam cracking is separated into two parts: a light hydrocarbon and heavy hydrocarbon reaction model (carbon number greater than five). The light hydrocarbon reaction model is generated using RMG [24]. RMG considers each species as unique and comprising a set of molecular structural isomers. When a reaction network is generated using RMG, it need not consider all the isomers in the real steam cracking process; instead, only a set of representative species are considered during the generation of a reaction network. The heavy reaction model is combined using different reaction models of pure compound feedstock and each reaction model is generated from the reaction families. The reaction coefficients can be obtained from the summary of experimental data (Table 1) in the work of Dente et al. [26]. The heavy hydrocarbon reaction model was generated using our own code.

The automatic generated reaction network may contain a large number of unimportant reactions and species. These reactions can increase the complexity of the model and make the model hard to solve. Thus, reaction model reduction is needed following the automatic model generation. The QSSA method is introduced first to remove the μ radicals in the reaction network [33]. As eq. (6) shows, the reaction rate of μ radicals can be treated as zero, based on the assumption.

Initiation reactions: unimolecular decomposition of C-C bonds			
CH_3-Csec	*Csec-Csec*	*Csec-Cter*	*Csec-Cquat*
5×10^{16}exp(-83,500/RT)	5×10^{16}exp(-81,000/RT)	5×10^{16}exp(-80,000/RT)	5×10^{16}exp(-78,000/RT)
H-abstraction reactions of alkyl radicals			
	Primary H-atom	Secondary H-atom	Tertiary H-atom
Primary radical	10^8exp(-13,500/RT)	10^8exp(-11,200/RT)	10^8exp(-9,000/RT)
Secondary radical	10^8exp(-14,500/RT)	10^8exp(-12,200/RT)	10^8exp(-10,000/RT)
Tertiary radical	10^8exp(-15,000/RT)	10^8exp(-12,700/RT)	10^8exp(-10,500/RT)
Isomerization reactions (Transfer of a primary H-atom)			
	1-4 H-transfer	1-5 H-transfer	1-6 H-transfer
Primary radical	10^{11}exp(-20,600/RT)	1.58×10^{10}exp(-14,500/RT)	3.16×10^9exp(-19,500/RT)
Secondary radical	10^{11}exp(-21,600/RT)	1.58×10^{10}exp(-15,500/RT)	3.16×10^9exp(-20,500/RT)
Tertiary radical	10^{11}exp(-22,100/RT)	1.58×10^{10}exp(-16,000/RT)	3.16×10^9exp(-21,000/RT)
Alkyl radical decomposition reactions (to form primary radicals)			
	Primary radical	Secondary radical	Tertiary radical
	10^{14}exp(-30,000/RT)	10^{14}exp(-31,000/RT)	10^{14}exp(-31,500/RT)
Corrections of decomposition rates to form			
Methyl radical	Secondary radical	Tertiary radical	Alkyl radical
exp(-2,500/RT)	exp(1,500/RT)	exp(2,500/RT)	$0.316\times$exp(8,000/RT)

Table 1. Reference kinetic parameters of pyrolysis reactions [26].

$$\frac{d[c]}{dt} = r_{formation} - r_{consumption} \approx 0 \tag{7}$$

Thus, eq. (8) can be derived from eq. (7):

http://dx.doi.org/10.5772/60558

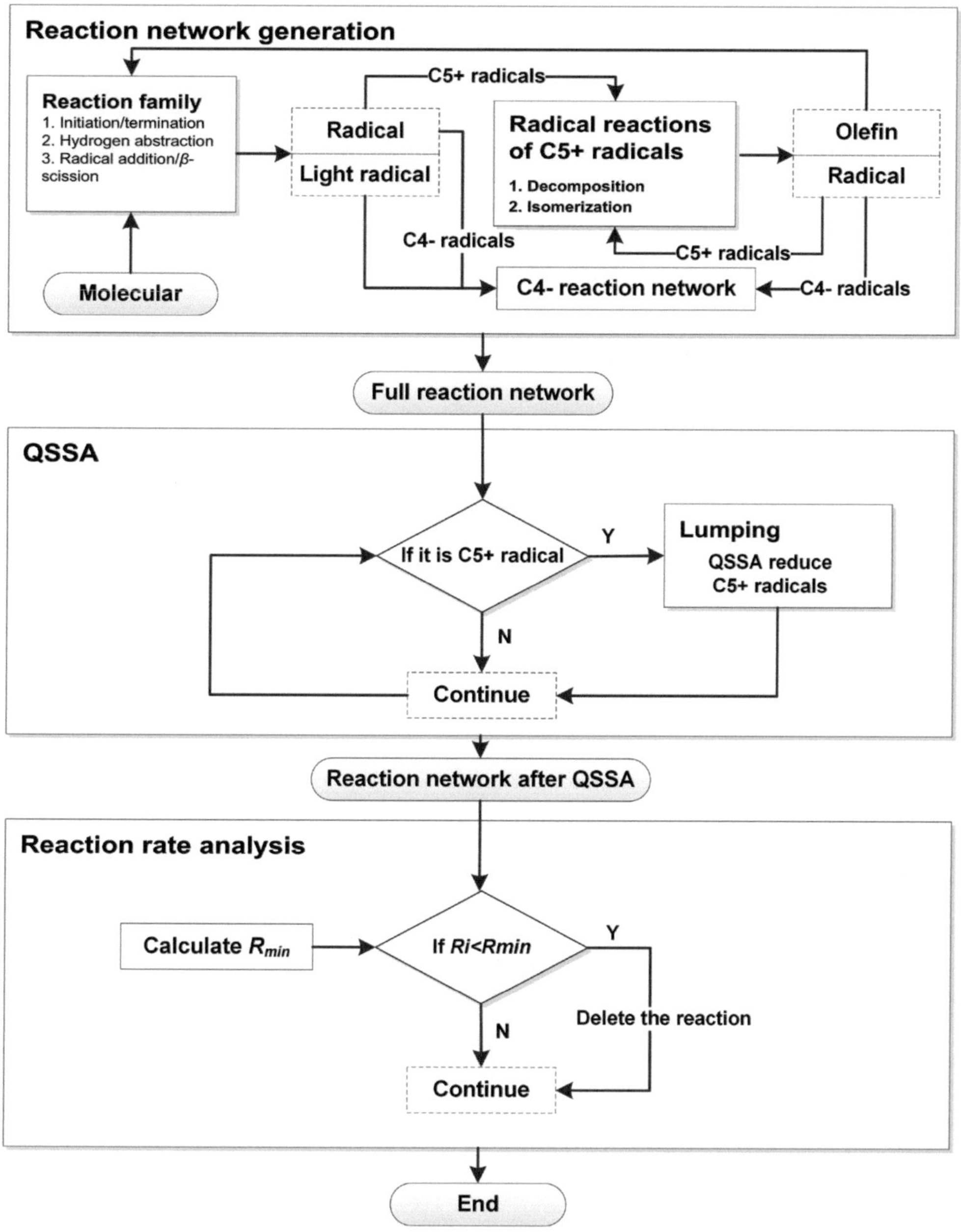

Figure 4. Flowchart of automatic generation and reduction of reaction network.

$$r_j + \sum_{i \neq j} k_{i,j} R_i - \left| \sum_{i \neq j} k_{j,i} + k_{d_i} \right| R_j = 0 \tag{8}$$

In eq. (8), r_j is the rate of direct formation of *j*-radical, $k_{i,j}$ is the rate constant for the isomerization reaction ($R_i \rightarrow R_j$) and k_{dj} is the total rate constant for the decomposition reactions of R_j; μ radicals are reduced using eq. (8). The number of species included in the model is decreased.

Reaction rate analysis is used to reduce the unimportant reactions in the reaction network. The average reaction rate in the reaction network reflects the importance of the reaction in the network. Thus, we can rank the reactions based on the average reaction rate and reduce the reactions where the reaction rate is less than R_{min} [44].

The flowchart of the automatic generation and reduction of the reaction network is shown in Figure 4.

d. Molecular-level model

The physical properties of some species (radical, non-common substances, etc.) involved in the model are difficult to obtain from databases. The physical properties of these species can be calculated using the group contribution method [42]. RMG also supplies a thermochemistry estimates utility using the group contribution method and was used in our model to automatically calculate the physical properties of these species.

Reaction	A	n	E (kcal/mol)	Reaction	A	n	E (kcal/mol)
CH3J+CH3J ↔ C2H6	8.26E+17	-1.4	1	BUT-2+C2H5J ↔ C4H7J +C2H6	3.36E+12	0	12.39
CH3J+C2H5J ↔ C3H8	1.37E+13	0	0	BUT-1+HJ ↔ C4H7J+H2	1.40E+04	2.36	1.11
C2H5J+C2H5J ↔ n-C4H10	1.15E+13	0	0	BUT-2+HJ ↔ C4H7J+H2	2.60E+06	2.38	2.8
C2H5J+HJ ↔ C2H6	1.00E+14	0	0	BUT-1+C3H7J ↔ C4H7J +C3H8	1.94E+02	2.96	6.79
C2H4+HJ ↔ C2H5J	1.55E+14	0	2.8	BUT-2+C3H7J ↔ C4H7J +C3H8	1.72E+12	0	12.29
C2H6+HJ ↔ C2H5J+H2	2.27E+08	1.75	7.51	BUT-1+C3H7J ↔ C4H7J +C3H8	3.12E-04	4.31	3.39
HJ+HJ ↔ H2	1.09E+11	0	1.5	BUT-2+C3H7J ↔ C4H7J +C3H8	3.36E+12	0	12.39
C3H7J+HJ ↔ C3H8	2.00E+13	0	0	BUT-1+C2H3J ↔ C4H7J +C2H4	5.11E+00	3.59	5.06
CH3J+C3H7J ↔ i-C4H10	6.64E+14	-0.57	0	BUT-2+C2H3J ↔ C4H7J +C2H4	4.64E+13	0	7.5
C3H8+C2H5J ↔ C3H7J+C2H6	3.08E+03	2.66	10.1	BUT-1+C4H7J ↔ C4H7J +BUT-1	3.12E-04	4.31	3.39
C3H8+HJ ↔ C3H7J+H2	1.60E+08	1.69	4.78	BUT-2+C4H7J ↔ C4H7J +BUT-1	3.36E+12	0	12.39
C3H6+HJ ↔ C3H7J	2.01E+13	0	2.1	C4H7J ↔ C4H7J	2.82E+08	1.28	27.89

http://dx.doi.org/10.5772/60558

Reaction	A	n	E (kcal/mol)	Reaction	A	n	E (kcal/mol)
C3H7J+HJ ↔ C3H8	1.00E+14	0	0	C4H7J+BUT-1 ↔ BUT-2+C4H7J	3.12E-04	4.31	3.39
CH3J+C3H7J ↔ n-C4H10	3.37E+13	0	0	C2H3J+C2H3J ↔ C4H6	7.23E+13	0	0
C2H4+CH3J ↔ C3H7J	1.94E+13	0	9	C4H6+HJ ↔ C4H7J	5.70E+13	0	4.3
C3H8+C2H5J ↔ C3H7J+C2H6	3.95E+03	2.71	12.91	C4H6+HJ ↔ C4H7J	7.42E+13	0	1
C3H8+HJ ↔ C3H7J+H2	3.77E+08	1.75	7.51	C3H5J+HJ ↔ C3H6	5.00E+13	0	0
C3H6+HJ ↔ C3H7J	1.18E+13	0	3.8	C3H5J+CH3J ↔ BUT-1	1.00E+13	0	0
C3H7J+C3H8 ↔ C3H8+C3H7J	3.08E+03	2.66	10.1	C3H6+CH3J ↔ C3H5J+CH4	2.41E+02	2.92	7.16
C3H7J ↔ C3H7J	1.94E+10	0.89	35.78	C3H6+C2H5J ↔ C3H5J +C2H6	1.01E-04	4.75	4.13
C2H6+CH3J ↔ C2H5J+CH4	1.67E+06	1.9	11.04	C3H6+HJ ↔ C3H5J+H2	1.30E+06	2.38	2.8
C3H8+CH3J ↔ C3H7J+CH4	1.67E+06	1.9	11.04	C3H6+C3H7J ↔ C3H5J +C3H8	8.61E+11	0	12.29
C3H8+CH3 ↔ C3H7J+CH4	2.90E+06	1.77	8.53	C3H6+C3H7J ↔ C3H5J +C3H8	1.01E-04	4.75	4.13
HJ+CH3J ↔ CH4	1.93E+14	0	0.27	C3H6+C2H3J ↔ C3H5J +C2H4	2.32E+13	0	7.5
HJ+CH4 ↔ H2+CH3J	1.22E+08	1.87	10.59	C3H6+C4H7 ↔ C3H5J +BUT-1	1.01E-04	4.75	4.13
C2H3J+HJ ↔ C2H4	5.36E+14	0	0.98	C3H5J+BUT-1 ↔ C3H6+C4H7J	3.12E-04	4.31	3.39
C2H3J+CH3J ↔ C3H6	7.23E+13	0	0	C3H5J+BUT-2 ↔ C3H6+C4H7J	3.47E+13	0	21.39
C2H3J+C2H5J ↔ BUT-1	1.00E+13	0	0	C4H7J+C2H3J ↔ C6H10	1.00E+13	0	0
C2H3J+CH4 ↔ C2H4+CH3J	5.36E+14	0	13.69	C3H5J+C3H5J ↔ C6H10	1.00E+13	0	0
C2H3J+C2H6 ↔ C2H4+C2H5J	5.30E+04	2.44	5.5	C4H6+CH3J ↔ C5H9J	1.46E+13	0	5.7
C2H3J+H2 ↔ C2H4+HJ	8.98E+12	0	10.3	C2H4+C3H5J ↔ C5H9J	1.43E+12	0	14.39
C2H3J+C3H8 ↔ C2H4+C3H7J	1.02E+03	3.1	8.82	C5H9 ↔ C5H9J	3.32E+08	1.1	29.39
C2H3J+C3H8 ↔ C2H4+C3H7J	9.30E+04	2.44	5.5	C4H9J+H ↔ n-C4H10	1.00E+14	0	0
C4H7J+HJ ↔ BUT-1	1.00E+14	0	0	n-C4H10+CH3 ↔ C4H9J +CH4	1.67E+06	1.9	11.04
BUT-1+CH3J ↔ C4H7J+CH4	8.34E+05	1.9	11.04	C4H9J+C2H6 ↔ n-C4H10+C2H5J	1.93E-05	5.28	7.78
C4H7J+C2H6 ↔ BUT-1+C2H5J	1.93E-05	5.28	7.78	C2H4+C2H5J ↔ C4H9J	3.98E+03	2.44	5.37
BUT-1+HJ ↔ C4H7J+H2	1.88E+08	1.75	7.51	n-C4H10+HJ ↔ C4H9J+H2	3.77E+08	1.75	7.51
C4H7J+C3H8 ↔ BUT-1+C3H7J	3.08E+03	2.66	10.1	BUT-1+HJ ↔ C4H9J	1.18E+13	0	3.8
C4H7J+C3H8 ↔ BUT-1+C3H7J	3.95E+03	2.71	12.91	C4H9J+C3H8 ↔ n-C4H10+C3H7J	3.08E+03	2.66	10.1
BUT-1+C2H3J ↔ C4H7J+C2H4	4.65E+04	2.44	5.5	C4H9J+C3H8 ↔ n-C4H10+C3H7J	3.95E+03	2.71	12.91

Reaction	A	n	E (kcal/mol)	Reaction	A	n	E (kcal/mol)
C2H4+C2H3J ↔ C4H7J	4.18E+10	0	5.2	n-C4H10+C2H3 ↔ C4H9J +C2H4	9.30E+04	2.44	5.5
C4H7J+HJ ↔ BUT-1	5.00E+13	0	0	n-C4H10+C4H7J ↔ C4H9J +BUT-1	3.95E+03	2.71	12.9
C4H7J+HJ ↔ BUT-2	5.00E+13	0	0	C4H9J+BUT-1 ↔ n-C4H10+C4H7J	3.12E-04	4.31	3.39
BUT-1+CH3J ↔ C4H7J+CH4	5.46E+13	0	10.39	C4H9J+BUT-2 ↔ n-C4H10+C4H7J	3.36E+12	0	12.39
BUT-2+CH3J ↔ C4H7J+CH4	4.82E+02	2.92	7.16	C4H9J+C3H6 ↔ n-C4H10+C3H5J	1.01E-04	4.75	4.13
BUT-1+C2H5J ↔ C4H7J+C2H6	3.12E-04	4.31	3.39				

Table 2. Elementary reaction network for light hydrocarbon steam cracking.

2.2. Case study: Naphtha steam cracking model simulation

A light hydrocarbon reaction model was generated using RMG, which contained 91 reactions and 26 species (Table 2). A naphtha steam cracking reaction network was generated based on the light hydrocarbon network. The naphtha reaction network contained 2424 reactions and 125 species.

Thirteen sets of industrial data (Table 3) were used here to verify the established multi-scale steam cracking model. The industrial data were taken from a steam cracking furnace designed by KBR (Kellogg Brown & Root). The data were collected twice a day and each set of data was the average value of these two parallel experiments as a means for preventing any errors. The multi-scale steam cracking simulation took roughly 70s CPU time. The simulation results and the industrial data of the mass fraction of the main products are shown in Figure 5, where the x-axis represents the industrial data and y-axis represents the simulation results. All points in Figure 5 are around the 1:1 diagonal line, which shows that the error of most results were small. Thus, the established multi-scale model was defined as accurate and robust; it satisfies industrial needs and can be applied further in our study of operating conditions optimization.

3. Multi-period steam cracking optimization

3.1. Multi-period optimization model for steam cracking

Coking is an unavoidable factor in the long-term steam cracking process. During the steam cracking process, coke forms on the inner walls of the tube. With coke formation, the internal diameter of the tube decreases, pressure drops increase and outer wall temperature increases. It is generally accepted that coke forms from unsaturated hydrocarbons and aromatics [43]; on this basis, an empirical model for coke formation rate is proposed, as shown in eq. (9).

$$r_c = \sum_{i=1}^{9} A_i \exp\left(-\frac{E_i}{RT}\right) c_i \qquad (9)$$

In eq. (9): 1=C_2H_2; 2=C_2H_4; 3=C_3H_6; 4=1-C_4H_8; 5=C_4H_6; 6=C_6H_6; 7=C_7H_8; 8=xylene; 9=C_2H_3-C_6H_5.

Steam cracking is a dynamic process, as coke grows inside the tube. However, coke formation is slow enough that we can divide the entire cracking period into a series of virtual steady state periods. Thus, the established multi-scale model can be used in each steady state period, with coke thickness updated between periods.

No.		1	2	3	4	5	6	7	8	9	10	11	12	13
Feed														
D15		0.7091	0.7106	0.7109	0.7099	0.7122	0.694	0.7099	0.6981	0.7109	0.6796	0.7121	0.7115	0.7109
M		94.1	94.76	94.68	94.29	94.31	94.43	94.6	93.67	94.74	95.26	95.26	95.41	94.83
T	°C	99.3	101.7	101.6	103.38	101.7	99.6	96.4	99	100.9	101.2	103.3	98.2	97
K		12.45	12.45	12.45	12.48	12.41	12.71	11.17	12.64	12.44	13.02	12.44	12.4	12.39
Operating conditions														
F	t/h	6.3686	5.413	5.4166	6.0792	5.1948	5.2302	4.9632	4.9812	5.3286	5.397	5.3726	5.431	5.5722
W/O	t/t	0.5	0.53	0.53	0.5	0.53	0.53	0.54	0.53	0.52	0.54	0.53	0.52	0.53
CIT	°C	639.6	639.4	639.5	633.4	640.9	639.8	642.5	641.4	636.5	634.8	638.8	640.8	639.1
COT	°C	871.6	867.6	867.2	872.6	865.6	869.2	865.8	861.6	865.4	863.4	868.4	865.8	867
CIP	Mpa	0.366	0.314	0.312	0.345	0.641	0.303	0.292	0.289	0.302	0.312	0.311	0.313	0.326
COP	Mpa	0.044	0.043	0.042	0.039	0.043	0.048	0.044	0.04	0.044	0.044	0.043	0.042	0.045
Yields of major products														
H_2	wt%	1.07	0.99	1.09	1.10	0.99	1.07	0.87	1.01	0.93	1.15	1.19	1.06	1.15
CH_4	wt%	16.55	15.62	16.73	16.83	15.39	16.73	13.90	15.76	15.15	18.09	18.80	16.98	17.64
C_2H_4	wt%	30.65	28.62	29.59	30.03	28.57	30.52	29.23	28.45	30.91	30.64	30.82	30.55	29.56
C_2H_6	wt%	3.79	3.44	3.54	3.56	3.34	3.64	3.49	3.69	3.61	3.63	3.65	3.68	3.60
C_3H_6	wt%	16.13	17.38	16.51	16.09	14.88	14.95	15.72	16.95	17.22	15.89	16.50	17.08	16.72
C_3H_8	wt%	0.90	1.74	0.87	0.82	0.83	0.69	0.99	0.88	1.53	0.96	1.05	1.21	1.17
C_4H_6	wt%	6.02	6.13	6.09	6.03	5.74	5.76	5.51	6.06	5.63	5.62	5.80	5.57	5.48
N-C_4H_8	wt%	1.41	1.59	1.70	1.54	1.40	1.31	1.46	2.07	1.38	1.37	1.43	1.58	1.57
I-C_4H_8	wt%	2.66	2.83	2.98	2.70	2.38	2.32	2.57	3.18	2.79	2.67	2.88	3.09	3.09
others	wt%	20.82	21.66	20.9	21.3	26.48	23.01	26.26	21.95	20.85	19.98	17.88	19.2	20.02

Table 3. Industrial data for feedstock, operating conditions and yields of major products.

Operation condition optimization is carried out based on the multi-period process model. Coil outlet temperature (COT) is selected as the variable to be optimized. Thus, the COT of all time periods of the multi-period model is discretized as [COT_1, COT_2,..., COT_n].

The optimization model is summarized below.

$$\max \frac{1}{N}\sum_{j=1}^{M}\sum_{i=1}^{N}\omega_t y_{j,t}^{o} \tag{10}$$

Subject to:

$$\frac{dy_{m,t}}{dL}=\frac{S_t}{V_t}\sum_i \nu_{i,m} r_{i,t} \quad \forall m\in\{1,2,\cdots,M\} \quad \forall t\in\{1,2,\cdots,N\} \tag{11}$$

$$\frac{dP_t}{dL}=-\frac{f\cdot E(L)\cdot G_t^2}{5.07\times\rho_t\cdot D_{i,t}\times 10^4} \quad \forall t\in\{1,2,\cdots,N\} \tag{12}$$

$$\frac{dT_t}{dL}=\frac{k_t\pi D_o\left(T_{w,t}-T_t\right)-\sum_m \Delta H_{fm,t}^{0}\cdot\frac{dy_{m,t}}{dL}}{\sum_m c_{pm,t}y_{m,t}} \quad \forall t\in\{1,2,\cdots,N\} \tag{13}$$

$$D_{i,t+1}=D_{i,t}-2\Delta\delta_{c,t}=D_{i,t}-2\cdot\frac{r_{c,t}}{\rho_{c,t}} \quad \forall t\in\{1,2,\cdots,N-1\} \tag{14}$$

$$r_{c,t}=\sum_{i=1}^{9}A_i\exp\left(-\frac{E_i}{RT}\right)c_{i,t} \quad \forall t\in\{1,2,\cdots,N\} \tag{15}$$

$$\max_t T_{w,t}\le T_M \quad \forall t\in\{1,2,\cdots,N\} \tag{16}$$

$$\left|COT_{t+1}-COT_t\right|\le\delta \quad \forall t\in\{1,2,\cdots,N-1\} \tag{17}$$

$$LB\le COT_t\le UB \quad \forall t\in\{1,2,\cdots,N\} \tag{18}$$

In objective function (10), M is the number of considered species, N is the period number, ω_t (t=1, 2,..., M) are weighted fractions based on the price of each species and $y^{o}_{j,t}$ is the mass fraction of selected species j in products of period t. Eqs. (11-13) describes mass balance, momentum balance and energy balance equations in period t (t=1, 2,..., N). In eq. (14), the internal diameter of period t+1 $D_{i,t+1}$ equals the internal diameter of the previous period $D_{i,t}$

minus the coke layer thickness. In eq. (16), peak outer-wall temperature should not exceed the maximum temperature of the tube material. Eq. (17) shows that the adjacent COT difference is restricted to a certain region to keep the operation stable. Eq. (18) shows the upper and lower boundaries of optimization variables.

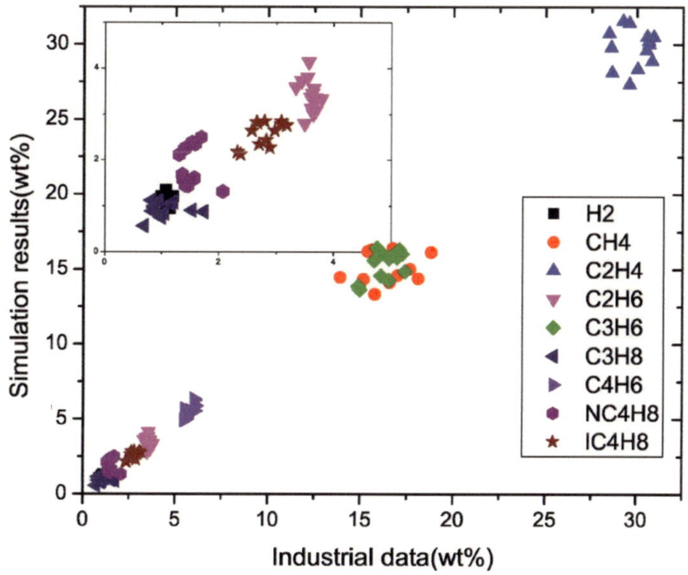

Figure 5. Simulation results compared with industrial data.

The optimization procedure is shown in Figure 6.

3.2. Surrogate coke thickness model

The most time-consuming part of the optimization is the simulation of the multi-period model. As Figure 6 shows, the only connection between adjacent periods is the coke thickness. Coke thickness is related to the feedstock component, furnace running time and operating conditions. As it has been assumed that the feedstock component is fixed between periods and only COT changes in operating conditions, as eq. (19) shows, then d_t is the coke thickness in period t.

$$d_t = f\left(\text{feed}, \text{period number}, COT_1, COT_2, \cdots, COT_t\right) \tag{19}$$

The accumulated coke thickness within a certain period δd_t is assumed to only be related to the furnace running time and COT_t (t=1, 2,..., N-1), shown as eqs. (20) and (21).

$$\begin{gathered}\delta d_t = d_{t+1} - d_t = f\left(t, COT_t\right) = a \times COT_t^2 + b \times t^2 + c \times COT_t + d \times COT_t + e \times t + f \\ \forall t \in \{1, 2, \cdots, N-1\}\end{gathered} \tag{20}$$

$$d_t = \sum_{t=1}^{i} \delta d_t = \sum_{t=1}^{i} f_t\left(t, COT_t\right) \tag{21}$$

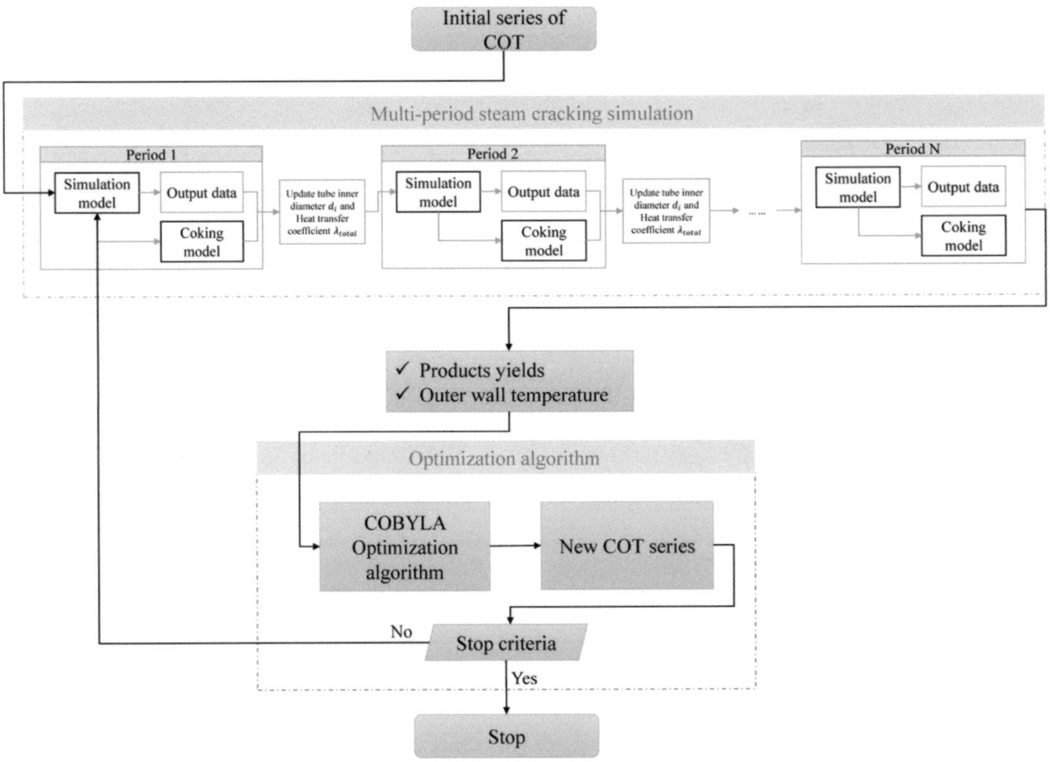

Figure 6. Optimization procedure with multi-period steam cracking simulation.

Thus, δd_t in period t can be regressed using the polynomial function shown as eq. (20). Here, the coke thickness data is generated using the original multi-period simulation model and based on this, a surrogate coke thickness model is obtained through regression. The coke thickness using the surrogate model and original multi-period simulation model are shown in Figure 7. Dots in Figure 7 are coke thickness from the original multi-period simulation model and the surface is from surrogate model.

The coke thickness from the surrogate model fits well with the original model; thus, the decoupled multi-period cracking model, combined with the surrogate coke thickness model was used in the multi-period simulation. The initial coke distribution along the serial operation periods was carried out using the surrogate model. Thus, the multi-period simulation problem was decoupled into N sub-problems and simulated, respectively, in parallel, as shown in Figure 8.

http://dx.doi.org/10.5772/60558

3.3. Optimization with parallel simulation

Figure 8 shows optimization with a parallel simulation procedure using the surrogate coke thickness model.

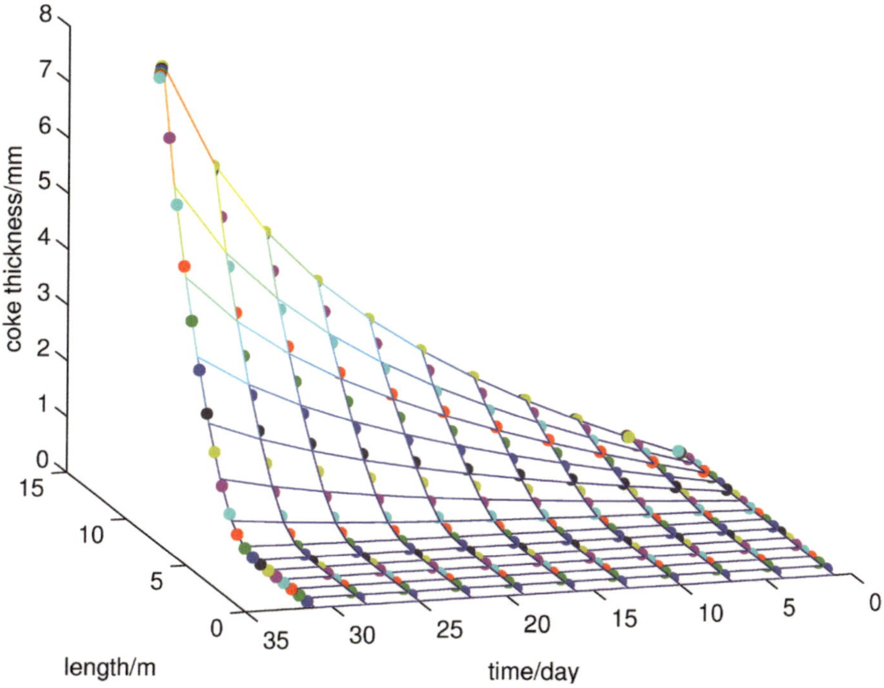

Figure 7. Coke thickness from the surrogate model and original multi-period model.

In Figure 8, a surrogate coke thickness model is regressed from the results of the original multi-period model and the surrogate coke thickness model generates the coke thickness distribution for each period prior to the simulation. Once the surrogate model is regressed, the process for each period can be simulated in parallel. The simulation results are sent to the optimization model. If the criteria are satisfied, the optimization stops; if not, the optimization model generates a new set of COT and returns it to the simulation model.

There are several common types of parallel computing methods: phase parallel, divide and conquer parallel, pipeline parallel, master-slave parallel and work pool parallel methods. In this instance, the work pool parallel job partitioning method was used in the optimization. Simulation for one period can be treated as one job in the parallel simulation. All the jobs are stored in the work pool; processors fetch jobs from the work pool as long as the work pool is not empty. A detailed scheme for parallel simulation is shown in Figure 9.

In Figure 9, *NP* is the resources that can be used in a parallel simulation (*NP*=computer number×processor number for each computer) and *NJ* is the period number in the multi-period simulation. If *NP*≥*NJ*, that means all jobs can be calculated in one iteration. Otherwise, *NP* jobs

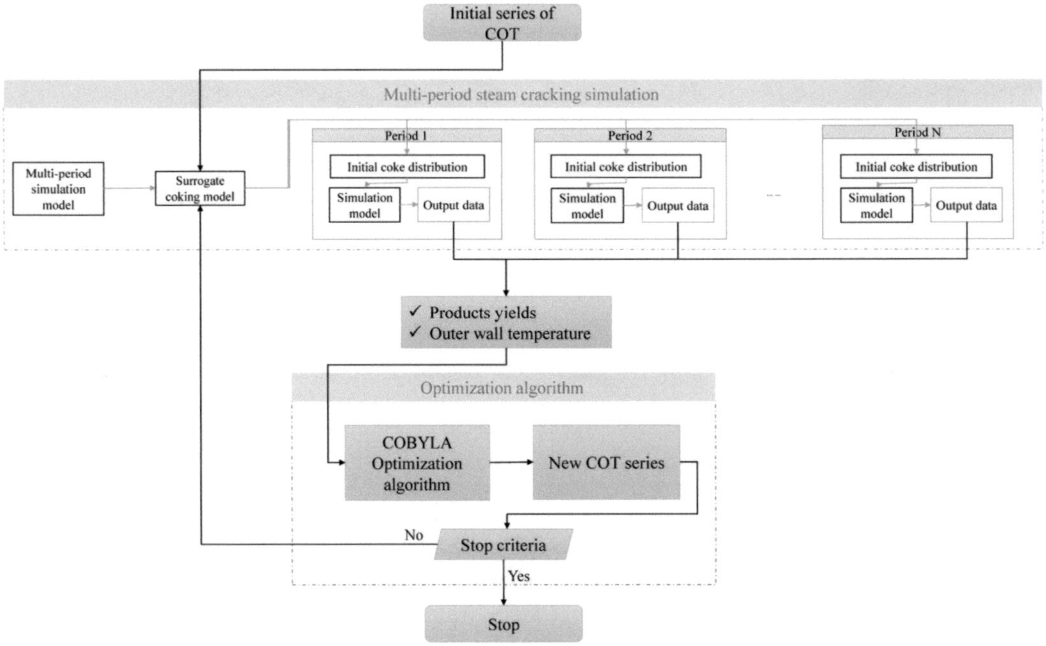

Figure 8. Optimization using a parallel simulation procedure.

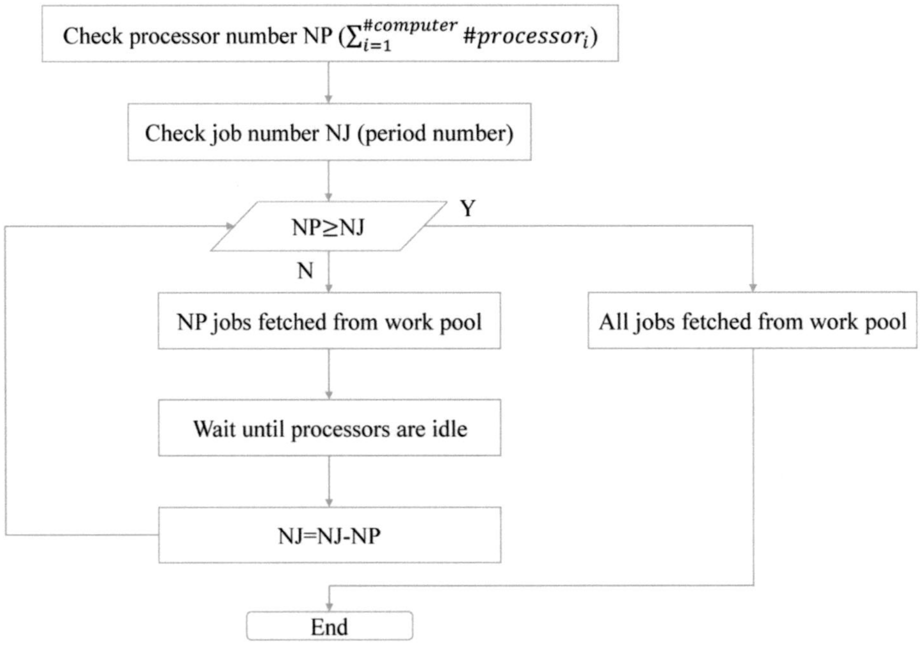

Figure 9. Detailed scheme for parallel simulation.

must be fetched from the work pool and be postponed until all processors are idle before the next iteration. Simulation will continue until the work pool is empty.

Another benefit of using parallel simulation is a warm start. To accelerate the steam cracking simulation, iteration information is stored after the first simulation, which is called a warm start. If the iteration information is recorded, it can be used as an initial value in the next simulation for reducing CPU time. A comparison between a warm start for serial and parallel simulation is shown in Figure 10.

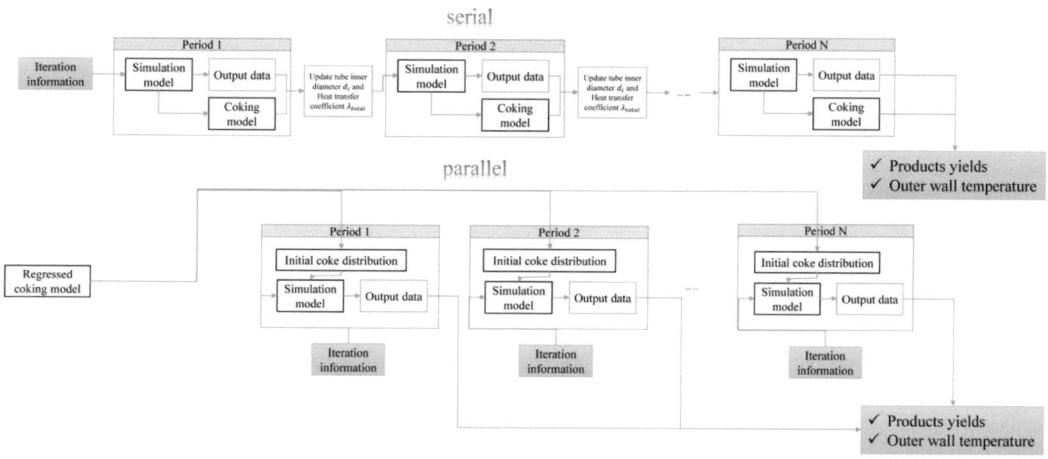

Figure 10. Comparison between a warm start for serial and parallel simulation, based on the multi-period steam cracking model.

Iteration information for the most recent period is stored in serial simulation. The iteration information for each period is not entirely the same. Inconsistency of iteration information makes the simulation slow. In a parallel simulation, however, iteration information for each period can be stored. This delivers a faster simulation for each period.

3.4. Case study: Naphtha steam cracking model optimization

An industrial multi-period steam cracking instance was used as a case study. Input data is shown in Table 4. The detailed feedstock composition was calculated using the Shannon entropy method [39]. An elementary reaction model and optimization with parallel simulation was used in this case study. The computer used in this optimization had eight cores of i7-2600 CPU and 4GB memory.

The optimization results show a 0.62% ethylene increase compared to the invariant operating condition that was implemented in the practice. The comparison of the serial optimization results and optimization with parallel simulation results are shown in Figure 11 and Figure 12. It can be seen that the tendency of serial and parallel optimization results were the same. There was a high outlet temperature and high conversion of the major products in the beginning and intermediate periods, and the outlet temperature of final periods were rather low for reducing the coke formation and for avoiding a high outer-wall temperature.

Feed							
Molecular weight	95.26						
H/C	2.11						
	P		I		N	A	
PINA	0.3096		0.3223		0.2856	0.0825	
	IBP	10%	30%	50%	70%	90%	EBP
ASTM (°C)	41.2	58.2	80.1	99.0	119.6	148.9	175.9
Operating condition							
Feed rate (t/h)	5.4						
Water/Oil (t/t)	0.54						
Coil inlet temperature (°C)	636.5						
Coil inlet pressure (Mpa)	0.31						
Coil outlet pressure (Mpa)	0.04						
Optimization variables							
Objective function	$\max \frac{1}{10}\left(0.52\sum_{t=1}^{10} y^o_{C_2H_4,t} + 0.48\sum_{t=1}^{10} y^o_{C_3H_6,t}\right)$						
Furnace runtime (day)	50						
NP	8						
T_M (°C)	1000.0						
δ (°C)	5.0						
LB (°C)	855.0						
UB (°C)	872.0						

Table 4. Supporting information for the steam cracking optimization model.

The error of parallel simulation compared to serial simulation is shown in Table 5.

The CPU time of serial optimization was 17.78hr, while the CPU time of parallel optimization was 2.08hr. There was a 8.55 x speedup compared with serial optimization, because parallel calculation and a warm start strategy were used. Thus, the operating conditions could be dynamically optimized to track the changing market conditions.

The error in optimization with the parallel simulation model was caused by the surrogate coke thickness model. In eq. (22),

$$\delta d_t = d_{t+1} - d_t = f\left(t, COT_t\right) \quad \forall t \in \{1,2,\cdots,N-1\} \tag{22}$$

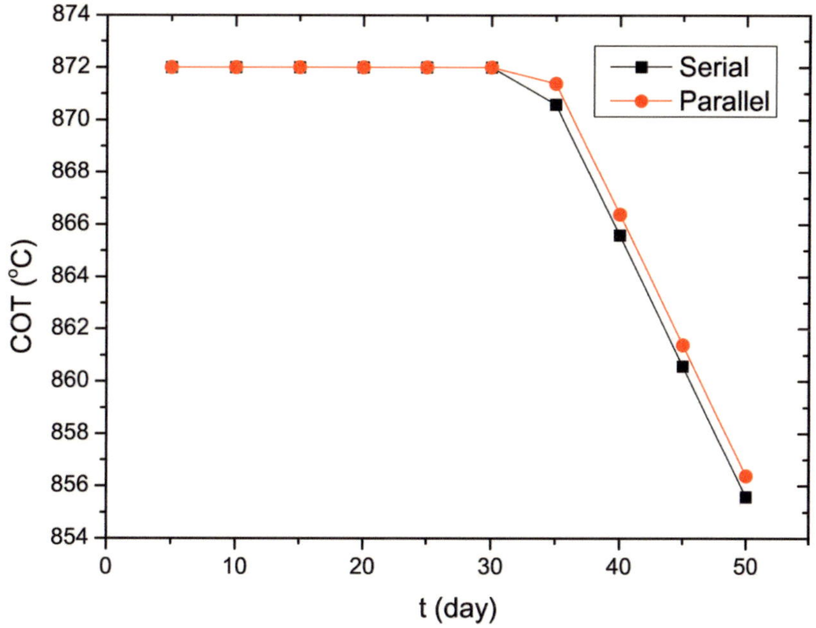

Figure 11. COT optimization results for serial and parallel simulation.

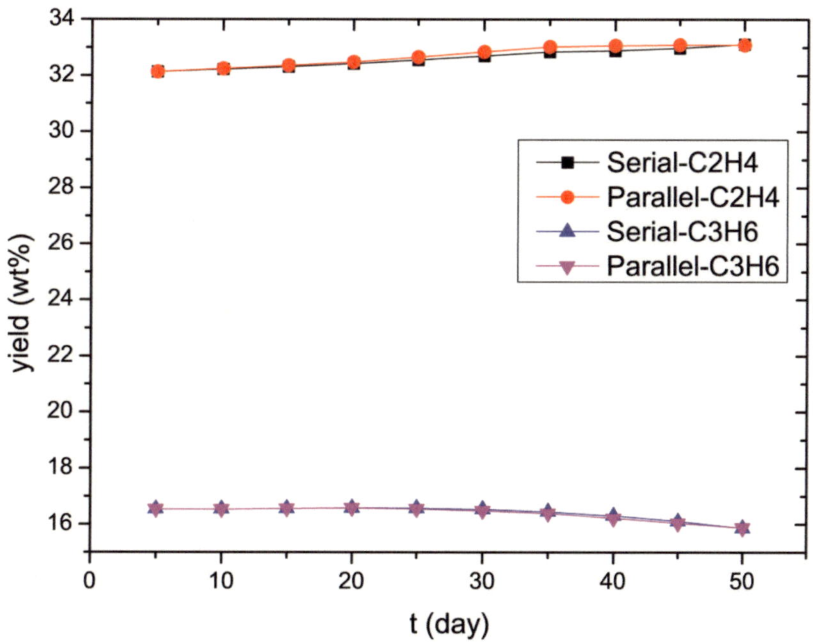

Figure 12. C_2H_4 and C_3H_6 optimization results for serial and parallel calculation.

The feedstock composition and operating conditions, except COT, were fixed. However,

$$\delta d_t = d_{t+1} - d_t = f\left(t, COT_1, COT_2, \cdots, COT_t\right) \quad \forall t \in \{1,2,\cdots,N-1\} \tag{23}$$

Accumulate coke thickness in each period was related to the operating conditions of all previous periods. Eq. (23) is simplified as eq. (22) to make the model easier to regress and the error caused by the simplification is acceptable according to the error shown in Table 5.

	COT			C_2H_4			C_3H_6		
Period	**Serial (°C)**	**Parallel (°C)**	**Error (%)**	**Serial (wt %)**	**Parallel (wt %)**	**Error (%)**	**Serial (wt %)**	**Parallel (wt %)**	**Error (%)**
1	872.00	872.00	0.00	32.13	32.13	0.01	16.54	16.54	0.00
2	872.00	872.00	0.00	32.22	32.24	0.05	16.55	16.55	0.00
3	872.00	872.00	0.00	32.31	32.36	0.14	16.57	16.57	0.01
4	872.00	872.00	0.00	32.42	32.48	0.21	16.59	16.59	0.02
5	872.00	872.00	0.00	32.56	32.65	0.27	16.58	16.56	0.14
6	872.00	872.00	0.00	32.70	32.84	0.44	16.55	16.49	0.34
7	870.58	871.38	0.09	32.84	33.02	0.54	16.46	16.40	0.34
8	865.58	866.38	0.09	32.89	33.07	0.54	16.31	16.22	0.52
9	860.58	861.38	0.09	32.98	33.09	0.35	16.11	16.04	0.39
10	855.58	856.38	0.09	33.11	33.09	0.04	15.87	15.88	0.07
Average error	-	-	0.04	-	-	0.26	-	-	0.18

Table 5. Summarized optimization results for serial and parallel calculation.

4. Conclusion

A multi-scale steam cracking model was established in this chapter. An elementary reaction network was generated to obtain a more accurate and robust model. The results showed that the established multi-scale model agrees well with the industrial data. A surrogate coke thickness model is thus proposed to make multi-period optimization with parallel simulation possible. A case study showed that multi-period optimization with parallel computing possesses an 8.55 x speedup compared to optimization with serial simulation. Parallel computing makes real-time optimization (RTO) possible in multi-period optimization. The average error of optimization using a parallel simulation model was 0.04% compared to optimization using a serial model. The error was deemed acceptable for the optimization of large-scale industrial steam cracking processes.

http://dx.doi.org/10.5772/60558

Optimization using the parallel computing method can also be used in other aspects of chemical processes. Due to the complexity of chemical processes, it is always difficult to conduct a simulation or optimization using a multi-scale model. The connections of different levels in a multi-scale model or different equipment in chemical process can be decoupled and transferred to a multiple sub-system model, which can be simulated in parallel. This method can bring powerful computing performance into play in chemical engineering.

Acknowledgements

The authors gratefully acknowledge the financial support of the National Basic Research Programme (No. 2012CB720500) and the NSFC (No. 21306100).

Author details

Lei Zhang, Hangzhou Wang, Tong Qiu and Bingzhen Chen*

*Address all correspondence to: dcecbz@tsinghua.edu.cn

Department of Chemical Engineering, Tsinghua University, Beijing, China

References

[1] Van Goethem, Marco et al. Equation-based SPYRO model and solver for the simulation of the steam cracking process. Computers & Chemical Engineering. 2001; 25(4): 905-911.

[2] Van Geem, K. Single event microkinetic model for steam cracking of hydrocarbons. Ph.D. thesis. Ghent University; 2006.

[3] Gao, X., Chen, B., He, X., Qiu, T., Li, J., Wang, C., Zhang, L. Multi-objective optimization for the periodic operation of the naphtha pyrolysis process using a new parallel hybrid algorithm combining NSGA-II with SQP. Computers & Chemical Engineering. 2008; *32*(11): 2801-2811.

[4] Yang, A., Marquardt, W. An ontological conceptualization of multiscale models. Computers & Chemical Engineering. 2009; 33(4): 822-837.

[5] Lucia, A. Multi-scale methods and complex processes: A survey and look ahead. Computers & chemical engineering. 2010; 34(9): 1467-1475.

[6] Davis, H. G., Farrell, T. J. Relative and absolute rates of decomposition of light paraffins under practical operating conditions. Industrial & Engineering Chemistry Process Design and Development. 1973; 12(2): 171-181.

[7] Sundaram, K. M., Froment, G. F. Modeling of thermal cracking kinetics-I: Thermal cracking of ethane, propane and their mixtures. Chemical Engineering Science. 1977; 32(6): 601-608.

[8] Sundaram, K. M., Froment, G. F. Modeling of thermal cracking kinetics-II: Cracking of iso-butane, of n-butane and of mixtures ethane-propane-n-butane. Chemical Engineering Science. 1977; 32(6): 609-617.

[9] Hirato, M., Yoshioka, S. Simulation of pyrolysis of naphtha, kerosene, and gas oil with a tubular reactor. International Chemical Engineering. 1973; 13(2): 347-355.

[10] Kumar, P., Kunzru, D. Modeling of naphtha pyrolysis. Industrial & Engineering Chemistry Process Design and Development. 1985; 24(3): 774-782.

[11] Rice, F. O. The thermal decomposition of organic compounds from the standpoint of free radicals. I. Saturated hydrocarbons. Journal of the American Chemical Society. 1931; 53(5): 1959-1972.

[12] Rice, F. O., Herzfeld, K. F. The thermal decomposition of organic compounds from the standpoint of free radicals. VI. The mechanism of some chain reactions. Journal of the American Chemical Society. 1934; 56(2): 284-289.

[13] Kossiakoff, A., Rice, F. O. Thermal Decomposition of Hydrocarbons, Resonance Stabilization and Isomerization of Free Radicals. Journal of the American Chemical Society. 1943; 65(4): 590-595.

[14] Gear, C. W. The automatic integration of ordinary differential equations. Communications of the ACM. 1971; 14(3): 176-179.

[15] Sundaram, K. M., Froment, G. F. Modeling of thermal cracking kinetics. 3. Radical mechanisms for the pyrolysis of simple paraffins, olefins, and their mixtures. Industrial & Engineering Chemistry Fundamentals. 1978; 17(3): 174-182.

[16] Scharfe, M., Ederer, H. J., Stabel, U., Ebert, K. H. Modeling of n-hexane pyrolysis: experimental investigations in a flow reactor at normal pressure. German chemical engineering. 1985; 8(2): 119-129.

[17] Joo, E., Park, S., Lee, M. Pyrolysis reaction mechanism for industrial naphtha cracking furnaces. Industrial & engineering chemistry research. 2001; 40(11): 2409-2415.

[18] Battin-Leclerc, F. Development of kinetic models for the formation and degradation of unsaturated hydrocarbons at high temperature. Physical Chemistry Chemical Physics. 2002; 4(11): 2072-2078.

[19] Warth, V., Battin-Leclerc, F., Fournet, R., Glaude, P. A., Côme, G. M., Scacchi, G. Computer based generation of reaction mechanisms for gas-phase oxidation. Computers & chemistry. 2000; 24(5): 541-560.

[20] Broadbelt, L. J., Stark, S. M., Klein, M. T. Computer generated pyrolysis modeling: on-the-fly generation of species, reactions, and rates. Industrial & Engineering Chemistry Research. 1994; 33(4): 790-799.

[21] Quann, R. J., Jaffe, S. B. Structure-oriented lumping: describing the chemistry of complex hydrocarbon mixtures. Industrial & engineering chemistry research. 1992; 31(11): 2483-2497.

[22] Di Maio, F. P., Lignola, P. G. KING, a kinetic network generator. Chemical engineering science. 1992; 47(9): 2713-2718.

[23] Hillewaert, L. P., Dierickx, J. L., Froment, G. F. Computer generation of reaction schemes and rate equations for thermal cracking. AIChE Journal. 1988; 34(1): 17-24.

[24] Song, J., Raman, S., Yu, J., Wijaya, C. D., Stephanopoulos, G., Green, W. H. RMG: the Next Generation of Automatic Chemical Reaction Mechanism Generator. proceedings. 2003. In AIChE Annual Meeting.

[25] Clymans, P. J., Froment, G. F. Computer-generation of reaction paths and rate equations in the thermal cracking of normal and branched paraffins. Computers & chemical engineering. 1984; 8(2): 137-142.

[26] Ranzi, E., Dente, M., Goldaniga, A., Bozzano, G., Faravelli, T. Lumping procedures in detailed kinetic modeling of gasification, pyrolysis, partial oxidation and combustion of hydrocarbon mixtures. Progress in Energy and Combustion Science. 2001; 27(1): 99-139.

[27] Turanyi, T., Berces, T., Vajda, S. Reaction rate analysis of complex kinetic systems. International Journal of Chemical Kinetics. 1989; 21(2): 83-99.

[28] Ravikeerthi, T., Thyagarajan, R., Kaisare, N. S., Aghalayam, P. Microkinetic model for NO–CO reaction: Model reduction. International Journal of Chemical Kinetics. 2012; 44(9): 577-585.

[29] Vajda, S., Valko, P., Turanyi, T. Principal component analysis of kinetic models. International Journal of Chemical Kinetics. 1985; 17(1): 55-81.

[30] Turanyi, T. Applications of sensitivity analysis to combustion chemistry. Reliability Engineering & System Safety. 1997; 57(1): 41-48.

[31] Bellamine, F. H., Elkamel, A. Model order reduction using neural network principal component analysis and generalized dimensional analysis. Engineering Computations. 2008; 25(5): 443-463.

[32] Lang, Y. D., Malacina, A., Biegler, L. T., Munteanu, S., Madsen, J. I., Zitney, S. E. Reduced Order Model Based on Principal Component Analysis for Process Simulation and Optimization. Energy & Fuels. 2009; 23(3): 1695-1706.

[33] Ranzi, E., Dente, M., Pierucci, S., Biardi, G. Initial product distributions from pyrolysis of normal and branched paraffins. Industrial & Engineering Chemistry Fundamentals. 1983; 22(1): 132-139.

[34] Abel, O., Birk, J. Real Time Optimization of Chemical Processes with Application to Olefins Production. AUTOMATISIERUNGSTECHNIK. 2002; 50(12): 586-596.

[35] Tarafder, A., Rangaiah, G. P., Ray, A. K. Multiobjective optimization of an industrial styrene monomer manufacturing process. Chemical Engineering Science. 2005; 60(2): 347-363.

[36] Li, C., Zhu, Q., Geng, Z. Multi-objective particle swarm optimization hybrid algorithm: An application on industrial cracking furnace. Industrial & engineering chemistry research. 2007; 46(11): 3602-3609.

[37] Klein, M. T., Hou, G., Bertolacini, R., Broadbelt, L. J., Kumar, A. Molecular modeling in heavy hydrocarbon conversions. CRC Press; 2010.

[38] Shannon, C. E. A mathematical theory of communication. ACM SIGMOBILE Mobile Computing and Communications Review. 2011; 5(1): 3-55.

[39] Zhang, L., Chen, B. Applications of Shannon's entropy theory to naphtha pyrolysis simulation. Chemical Engineering & Technology. 2012; 35(2): 281-286.

[40] Jae Lee, W., Froment, G. F. Ethylbenzene dehydrogenation into styrene: kinetic modeling and reactor simulation. Industrial & Engineering Chemistry Research. 2008; 47(23): 9183-9194.

[41] Xu Q., Chen, B., He, X. A fast simulation algorithm for industrial cracking furnaces. Hydrocarbon processing. 2002; 81: 65-68.

[42] Constantinou, L., Gani, R. New group contribution method for estimating properties of pure compounds. AIChE Journal. 1994; 40(10): 1697-1710.

[43] Kumar, P., Kunzru, D. Coke formation during naphtha pyrolysis in a tubular reactor. The Canadian Journal of Chemical Engineering. 1987; 65(2): 280-285.

[44] Zhang, L. Radical Reaction Model Based Simulation and Optimization Method for Steam Cracking Process. Ph.D. thesis. Tsinghua University; 2014.